BATTERY SYSTEMS ENGINEERING

BATTERY SYSTEMS ENGINEERING

Christopher D. Rahn and Chao-Yang Wang

The Pennsylvania State University, USA

A John Wiley & Sons, Ltd., Publication

This edition first published 2013
© 2013 John Wiley & Sons, Ltd

Registered office
John Wiley & Sons Ltd, The Atrium, Southern Gate, Chichester, West Sussex, PO19 8SQ, United Kingdom

For details of our global editorial offices, for customer services and for information about how to apply for permission to reuse the copyright material in this book please see our website at www.wiley.com.

The right of the author to be identified as the author of this work has been asserted in accordance with the Copyright, Designs and Patents Act 1988.

Wiley also publishes its books in a variety of electronic formats. Some content that appears in print may not be available in electronic books.

Library of Congress Cataloging-in-Publication Data applied for.

ISBN: 9781119979500

Typeset in 10/12pt Times by Aptara Inc., New Delhi, India

To our parents

Contents

Preface

Energy storage is a critical and growing need in the drive to increase the efficiency and effectiveness of power systems. In the quest for higher fuel efficiency, energy storage is becoming increasingly important in ground transportation. Hybrid electric vehicles (HEVs) that recover the energy otherwise dissipated during braking are commanding a growing share of the passenger car, truck, and bus markets. Electric vehicles and plug-in HEVs charge using low-cost energy from the grid. Renewable energy sources such as wind and solar require energy storage to buffer power production deficits. Home energy storage can reduce costs by taking grid power during low-demand periods (e.g., at night) and reducing grid power during high-demand periods.

There are many ways to store energy (e.g., flywheels, ultra-capacitors, and compressed air) but batteries are the best choice for most applications. Batteries can be scaled from small (cell phone), to medium (HEVs), to large (grid) applications. They are highly efficient and have high energy-to-weight ratios. There are safe and recyclable designs. Cost and battery life, however, are concerns that prevent more widespread application of batteries for energy storage applications. Researchers are continually inventing lower cost and longer life battery chemistries. Efficient and life-extending battery management systems, designed using the techniques described in this book, can also address these concerns.

The dynamic environment of many energy storage applications requires battery management systems that are more advanced than would be required for a typical battery-powered device (e.g., laptop or cell phone). Simple battery-powered devices only require charging at periodic intervals and then draw low current, slowly discharging the pack until it is time to recharge again. HEVs, on the other hand, require fast and high-current energy storage associated with dynamic acceleration and braking of the vehicle. This rapid charge–discharge cycling of the battery pack requires sophisticated battery management systems to regulate the current in and out of the pack in real time. An effective battery management system sets the current limits low enough to maximize the battery life and ensure safety but high enough to maximize power output.

Battery systems engineering sits at the crossroads of chemistry, dynamic modeling, and systems/controls engineering, requiring a multidisciplinary approach. Battery chemists/ engineers understand the electrochemistry and materials issues required to design batteries but may not have the background to address the complex mathematical modeling and control systems design required for efficient battery management algorithms. Mathematical modelers may be able to develop accurate models of battery cells but these models are often not easily adopted for systems engineering owing to the complexity of the underlying partial differential equations. Systems engineers have the controls and dynamics background to analyze, design, and simulate the system response but may not understand the underlying chemistry or modeling.

This book aims to develop the multidisciplinary area of battery systems engineering by providing the background, models, solution techniques, and systems theory that are necessary for the development of advanced battery management systems. Systems engineers in chemical, mechanical, electrical, or aerospace engineering who are interested in learning more about advanced battery systems will benefit from this text. Chemists, material scientists, and mathematical modelers can also benefit by learning how their expertise affects battery management. The book could be used in an advanced undergraduate technical elective course or for graduate-level courses in engineering.

We would like thank our students, post-doctoral scholars, and research associates for their help in the preparation of this book. In particular, Kandler Smith, Yancheng Zhang, Ying Shi, Githin Prasad, and Zheng Shen have made significant contributions to the text and deserve our thanks. Students who took the first two offerings of the course *Battery Systems Engineering* at Penn State have also provided comments and corrected typos, including Kelsey Hatzell, Ed Simoncek, Ryan Weichel, and Tanvir Tanim. Chao-Yang gratefully acknowledges his wife, May M. Lin, and daughters, Helen and Emily, for their constant love, support, and strength. I am likewise grateful for the love, support, and encouragement of my wife Jeanne, daughter Katelin, and sons Kevin and Matthew.

Christopher D. Rahn
Chao-Yang Wang

1

Introduction

High energy costs drive the development of power systems with increased efficiency and effectiveness. One way to increase performance is to store energy that cannot be used at the time of its production. Batteries are being used in hybrid vehicles and renewable energy applications for this purpose. These applications can require dynamic cycling of the battery that can lead to poor performance and premature aging if not controlled by a sophisticated battery management system (BMS). BMSs that are based on accurate system models hold great promise for extending the life and increasing the performance of energy storage systems. This chapter motivates the need for model-based battery system engineering and introduces the electrochemistry and design of battery cells and packs.

1.1 Energy Storage Applications

Energy storage is vitally important to many applications, ranging from small-scale portable electronics to large-scale renewable energy sources. Portable electronic devices that use batteries include video/audio players, medical equipment, power tools, meters and data loggers, and remote sensors [1]. In these applications, batteries free the user from power cords and enable portable use. The batteries in these devices are discharged over time and then recharged periodically. Energy storage can also be used in large-scale applications to reduce oil, gas, and coal consumption. Hybrid vehicles for ground transportation and renewable (e.g., wind and solar) energy sources make use of batteries to store energy that cannot be used at the time of its production. The charge and discharge cycles in these applications are more frequent and dictated by the variable power supply and demand.

To increase the fuel efficiency of ground vehicles, batteries are being used to supplement and sometimes replace the power provided by liquid fuel. Figure 1.1 shows four pioneering vehicles that use batteries to increase fuel efficiency and performance. The Toyota Prius in Figure 1.1a is a hybrid electric vehicle (HEV). It uses a nickel–metal hydride (Ni–MH) battery pack manufactured by Panasonic. The Nissan Leaf and Tesla Roadster in Figure 1.1b and d, respectively, are electric vehicles (EVs). The Leaf uses a laminated lithium-ion (Li-ion) battery pack developed by Nissan–NEC and the Tesla uses a specially built pack with thousands of

Battery Systems Engineering, First Edition. Christopher D. Rahn and Chao-Yang Wang.
© 2013 John Wiley & Sons, Ltd. Published 2013 by John Wiley & Sons, Ltd.

Figure 1.1   Pioneering hybrid vehicles: (a) Toyota Prius (© Toyota). (b) Nissan Leaf (© 2012, Nissan. Nissan, Nissan model names and the Nissan logo are registered trademarks of Nissan). (c) Chevrolet Volt (photo taken by US National Highway Traffic Safety Administration). (d) Tesla Roadster (© Tesla Motors, Inc.)

18650 (18 mm diameter and 65 mm long) Li-ion cells. The Chevy Volt in Figure 1.1c is a plug-in HEV (PHEV) or extended-range electric vehicle (EREV) that has a Li–polymer battery pack supplied by LG-Chem.

HEVs are commanding a growing share of the passenger car, truck, and bus markets. Hybrid powertrains consist of an internal combustion engine (ICE), powertrain, electric motor, and batteries. HEVs conserve energy because they have the ability to:

1. **Eliminate engine idling.** The engine stops when the vehicle is stationary.
2. **Recover and store energy.** The electric motor is used as a generator to brake the vehicle. The regenerated energy is stored in the batteries.
3. **Boost power.** The electric motor and engine work together to increase torque during acceleration.
4. **Operate efficiently.** The engine can be run at its most efficient speed and the electric motor can provide power during off-peak operation.

HEVs vary in cost and complexity from simple retrofits to complete redesigns of existing ICE vehicles. Micro hybrids use a higher power starter/alternator to provide the advantages of eliminating engine idling. Soft hybrids add some regenerative braking and low-speed movement under electric power. Mild hybrids insert an electric motor/generator into the drive axle to provide all of the benefits of hybrid operation. The parallel drive train often used in mild hybrids allows the electric motor/generator to run the vehicle and boost power at low speeds. Full hybrids often use a series/parallel drive train that has all of the benefits of the parallel drive train. They can be used to decouple the motor speed from the vehicle speed so that the motor can run more often at peak efficiency. Full hybrids are the most efficient and complicated HEVs, with the batteries carrying a larger percentage of the load, continually being charged and discharged.

The battery packs in PHEVs charge directly from the electric grid and run the vehicle for a distance in pure electric mode with zero gas consumption and emissions. The vehicle also has an ICE that can be used to extend the electric-only range or increase the speed above the electric-only limit. After the batteries have been depleted to a specified level, the vehicle operates in full hybrid mode until it can be fully recharged from the grid. The Chevrolet Volt PHEV uses a variation on the series drivetrain where the engine drives a generator and is not mechanically connected to the drive wheels. A series drivetrain cannot use the engine and electric motor simultaneously to provide a power boost for quick acceleration.

EVs are zero-emission vehicles that charge from the grid. Batteries provide all of the power and energy for the drive motor. The key consideration in the design of an EV is the weight and cost of the battery pack. Lighter weight batteries typically cost more. The batteries are charged and then slowly discharged during operation, with regenerative braking providing intermittent recharge pulses.

The charging infrastructure required for EVs is a major challenge to the widespread adoption of this technology. Chargers at home or work can take hours to charge the battery for an EV or PHEV without too much inconvenience to the driver. If an EV is on the road and needs a quick charge, however, the infrastructure for fast (5 min) charging should be widely available. The charging power for a 5 min charge is 12 times the power that the pack can provide for 1 h. Long-range (300 mi) EVs require roughly a 75 kWh pack, so a 5 min charge would require 0.9 MW from the grid. As more and more EVs with longer and longer ranges replace gas-powered vehicles, the power grid infrastructure will need to drastically increase to accommodate the increased demand.

Passenger cars make up the bulk of the HEV market, but trucks and buses have also been converted to HEVs and EVs. Figure 1.2 shows, for example, an all-electric switchyard locomotive developed by Norfolk Southern. The locomotive is charged during the night and then is used for an 8 h shift, moving freight cars around the yard to form trains. Over 1000 lead–acid (Pb–acid) batteries are used to power the electric traction motors.

Renewable energy sources such as wind and solar and smart-grid technology require energy storage to buffer power production deficits. Wind and solar energy sources do not produce energy at a continuous rate. Energy produced in excess of demand can be stored in large-scale battery farms to be used at a later time. Home energy storage can reduce costs by taking grid power during low-demand periods (e.g., at night) and reducing grid power during high-demand periods. A smart grid regulates the power delivered to individual homes so that household energy storage can bridge the power gaps.

1.2   The Role of Batteries

There are many ways to store energy (e.g., flywheels, ultra-capacitors, and compressed air), but batteries are the best choice for most applications. Batteries can be scaled from small (cell phone), to medium (HEVs), to large (grid) applications. They are highly efficient and have high energy-to-weight ratios. They are safe and often recyclable. Cost and battery life, however, are concerns that prevent more widespread application of batteries for energy

Figure 1.2 Norfolk Southern Electric switchyard locomotive, NS999 (photo courtesy of Norfolk Southern Corp.)

storage applications. Researchers are continually inventing lower cost and longer life battery chemistries. As batteries become integral parts of high-volume products, economies of scale will reduce costs. A life-extending BMS, designed using the techniques described in this book, ensures that the battery pack is being used in a most efficient and cost-effective manner.

1.3 Battery Systems Engineering

Battery systems engineering sits at the crossroads of chemistry, dynamic modeling, and systems engineering. Battery chemists/engineers understand the electrochemistry and materials issues required to design batteries but may not have the background to address the complex mathematical modeling and control systems design associated with efficient battery management algorithms. Mathematical modelers can develop accurate models of battery cells but these models are often not easily adopted for systems engineering. Systems engineers have the controls and dynamics background to analyze, design, and simulate the system response but may not understand the underlying chemistry or models.

One of the main objectives of this book is to bring batteries into the realm of systems engineering. From a systems engineering perspective, battery packs are multi-input, multi-output systems. The primary input, current, is prescribed by the supply and demand from the powered device. The primary output is the battery voltage. Other outputs include temperature,

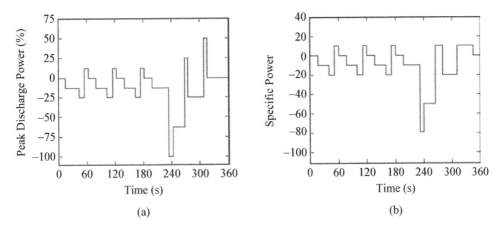

Figure 1.3   Battery cycling profiles for HEVs: (a) dynamic stress test (DST) and (b) simplified federal urban driving schedule (SFUDS)

individual battery or cell voltages, and ionic concentration distributions within a given cell. Systems engineers need cell, battery, and pack models in standard (e.g., state variable and transfer function) forms that can be used to predict, estimate, and control these outputs.

The dynamic environment of many energy storage applications requires advanced BMSs. BMSs are often concerned with charging protocols because applications require fully charging the pack at periodic intervals. The battery-powered device (e.g., laptop) then draws low current, slowly discharging the pack until it is time to recharge again. An HEV, on the other hand, requires fast and high-current energy storage associated with dynamic acceleration and braking of the vehicle. Figure 1.3 shows, for example, two HEV battery cycling profiles. The power into and out of the battery pack changes quickly over the 6 min cycles. This rapid charge–discharge cycling of the battery pack requires sophisticated BMSs to regulate the current in and out of the pack in real time. An effective BMS sets the current limits low enough to maximize the battery life and ensure safety but high enough to maximize power output.

Figure 1.4 shows a schematic diagram of the electromechanical system of an HEV. The battery system consists of cells grouped into modules that make up the battery pack, the BMS, and the thermal management system. The power electronics interface the battery system to the motor/generator that is mechanically coupled to an ICE through a transmission. The power electronics typically include high-power switching circuits, inverters, DC–DC converters, and chargers. The transmission either connects both the motor/generator and the engine to the wheels (parallel configuration), only the motor/generator to the wheels (series configuration), or some combination of the two (hybrid configuration).

While there are significant challenges in the development of new battery chemistries, power electronics, and motor/generators for HEV/PHEV/EV application, the focus of this book is on the dynamics of commercially available cells/packs and the development of estimation/control software that runs on-board the vehicle. The dynamic models can be used to simulate and optimize the system response. The software is based on the developed models and predicts and controls the battery-pack response to optimize performance and long pack life. Batteries

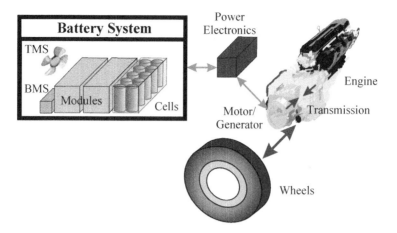

Figure 1.4   PHEV electromechanical system schematic

are the highest cost item in HEV, PHEV, and EV powertrains, so their optimal utilization is paramount to the development of affordable vehicles.

1.4   A Model-Based Approach

Batteries can be designed using empirical or model-based approaches. In an empirical approach, cells are built and tested for performance. Based on the results of the tests, the batteries are redesigned and tested again. This is a time-consuming and expensive process. In a model-based approach, a model is used to predict performance based on the battery design. This process is termed computer-aided engineering (CAE) because the battery can be designed and optimized relatively quickly on a computer. Model-based design ensures that the batteries developed have the highest possible performance, making them competitive in the marketplace.

A model-based approach builds upon a fundamental physics-based model that predicts the battery response. The model starts with the electrochemical and physical partial differential equations (PDEs) that govern the flow of ions through a battery cell. The model requires knowledge of geometric parameters (e.g., lengths, areas) that can be independently measured, physical constants (e.g., Faraday constant), and parameters that may not be independently measurable and/or known (e.g., diffusion coefficients). Given a time-varying battery input current, the model predicts the battery time response, including output voltage. The best models have parameters that are all measured independently and performance that closely matches experiments. The unknown parameters in a model provide extra knobs for the modeler to adjust to get good agreement with the experimental data. The process of model validation includes testing the model under a variety of inputs and minimizing the error between the model-predicted and experimental responses. Once the model has been validated, the input parameters can be varied according to different battery designs and the performance predicted. Thus, the battery can be optimized for maximal performance.

BMSs can also be designed using empirical or model-based approaches. Almost all BMSs rely on battery models, but the sophistication varies considerably. At the lowest level, heuristic models that roughly predict the observed performance are used. More advanced empirical models that fit equivalent circuits to the measured response over a specified frequency bandwidth have been applied extensively. The most advanced BMSs, however, are based on fundamental models of the batteries. These models are more difficult to derive and simplify for real-time applications, but they are based on the underlying physics and electrochemistry of the battery. The relationships between the response and system parameters are known. Fundamental model-based controllers have a built-in understanding of the underlying processes, allowing them to be more efficient, accurate, and safe.

1.5 Electrochemical Fundamentals

Figure 1.5 shows a schematic diagram of a battery cell. It consists of positive and negative electrodes immersed in an electrolyte solution. The electrodes can be solid material or porous to allow the electrolyte to infiltrate through. The separator prevents electrons from flowing but allows positive and negative ions to migrate between the two electrodes through the electrolyte. The positive and negative current collectors provide a pathway for electrons to flow through an external circuit. During discharge, the negative electrode is the anode and the positive electrode is the cathode. Positive ions move from the anode to the cathode through the electrolyte and separator. Negative ions move in the opposite direction. The anode builds up negative charge and the cathode builds up positive charge, creating the cell voltage $V(t)$. Negatively charged electrons flow through an external load from the anode to the cathode, creating a current in the opposite direction. The sign convention for positive current is in the opposite direction of the

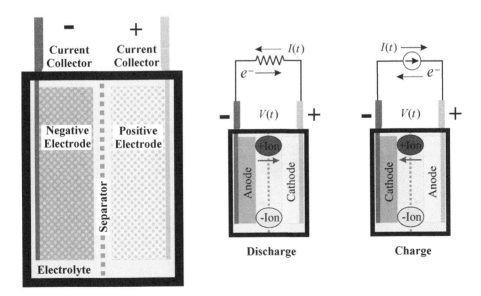

Figure 1.5 Simple cell under discharge and charge

electron flow. During charge, the process is reversed and electrons are forced into the cathode (now the negative electrode).

During charging, the negative electrode material dissolves in the electrolyte solution to form a positive ion and an electron in what is called an oxidation reaction. The positive electrode consumes electrons by depositing positive ions from the electrolyte in what is called a reduction reaction. The reactions are reversible in secondary (or rechargeable) batteries so that discharging the batteries returns the electrodes to their pre-charged states. The ions move through the electrolyte under diffusion and migration. Diffusion results from the existence of a concentration gradient in the electrolyte. Over time, if there is no ion production, the ions in the electrolyte diffuse evenly throughout the cell. Migration results from the presence of the electric field generated by the positive and negative electrodes. The positive ions migrate toward the negative electrode and the negative ions migrate toward the positive electrode. The movement of ions through the electrolyte and electrons through the external circuit enable the storage and release of energy.

1.6 Battery Design

Batteries come in all shapes and sizes, but the most common form factors are either prismatic (generally a rectangular prism) or cylindrical. Figure 1.6 shows Pb–acid and Ni–MH batteries and battery packs from Panasonic. The valve-regulated lead–acid (VRLA) batteries are prismatic and the Ni–MH batteries are manufactured in both cylindrical and prismatic form factors. The HEV battery pack shown in Figure 1.6(c) is made from many prismatic Ni–MH batteries.

VRLA batteries are typical of what one sees in ICE vehicles for starting, lighting, and ignition. Pb–acid cells produce around 2 V, so the batteries consist of several cells in series to produce the desired voltage of, for example, 6 V (three cells) or 12 V (six cells). A fully charged 12 V VRLA battery, however, can produce almost 15 V and be discharged to 8–10 V. The battery consists of lead plates and separators immersed in a diluted sulfuric acid electrolyte. Alternating plates of Pb and PbO_2 form the negative and positive electrodes, respectively. The current (and power) of the battery is proportional to the plate area. The battery case has a vent that opens if the internal pressure builds up to a sufficiently high level due to extreme overcharge conditions.

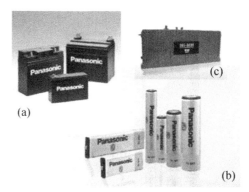

Figure 1.6 Panasonic batteries: (a) VRLA, (b) Ni–MH, and (c) Ni–MH pack (© Panasonic)

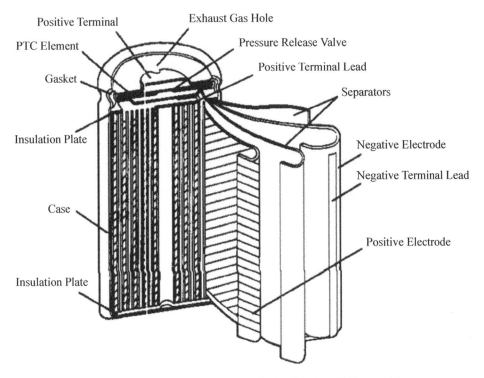

Positive Terminal
Exhaust Gas Hole
PTC Element
Pressure Release Valve
Positive Terminal Lead
Gasket
Separators
Insulation Plate
Negative Electrode
Negative Terminal Lead
Case
Positive Electrode
Insulation Plate

Figure 1.7   Panasonic Li-ion cylindrical cell design (© Panasonic)

The design of a Panasonic Li-ion cylindrical cell is shown in Figure 1.7. The battery is fabricated from four layers of material that are rolled up to form a cylinder. The layers are the positive electrode, separator, negative electrode, and then a second separator. The second separator layer keeps the positive and negative electrodes apart in the rolled configuration. Leads connect the positive electrode to the top terminal and the negative electrode to the bottom terminal. Li-ion batteries have a nominal voltage of over 3 V. To form higher voltage batteries, the cylindrical cells are stacked in series and sealed together. Higher current can be obtained by increasing the electrode area, resulting in a larger diameter or longer length cell.

The sealed prismatic Ni–MH cells shown in Figure 1.6b also have the same layered structure as cylindrical cells, but the layers are not rolled up. These layers can be stacked to increase the battery voltage from the nominal Ni–MH cell voltage of around 1 V. Prismatic cells are often easier to integrate into HEV packs like the one shown in Figure 1.6c. Packs connect individual batteries in series and parallel to raise the voltage and current to the desired values, respectively.

1.7   Objectives of this Book

The main objective of this book is to provide the framework for battery systems engineering as a viable field of study. The importance of batteries in energy consumption and production is growing. Batteries are often the most expensive and least well understood parts of these complex systems. This book targets design engineers who are not sufficiently familiar with batteries to be able to analyze, integrate, and optimize them as part of a more complicated system. We intend to provide a self-contained, fundamental approach to the modeling, analysis, and design of battery systems that places them within the same framework of mechanical, electrical, fluid, thermal, and computer models that engineers use to design complex mechatronic systems like HEVs and renewable energy plants.

To achieve this objective, we first develop battery models that can be understood and used by systems engineers with limited electrochemistry backgrounds. These models are well known for the Pb–acid, Ni–MH, and Li-ion chemistries discussed in Chapter 2. New battery chemistries that are developed in the future will undoubtedly use the same building blocks of conservation laws and reaction kinetics that provide the governing equations presented in Chapter 3. The focus is not on electrochemistry and the derivation of the governing equations, but on how to convert these distributed parameter models to standard forms that are commonly used by systems engineers. In Chapter 4 we study methods of spatially discretizing the underlying PDEs to reduce them to ordinary differential equations with one independent variable: time. These state-variable models are well known to systems engineers and form the basis of mechatronic systems analysis, design, and control. These models are then simulated in Chapter 5 to predict the charge–discharge, cycle, and frequency response. In Chapter 6, complete models of Pb–acid, Ni–MH, and Li-ion cells are presented and simulated using the techniques developed in Chapters 2–5.

Second, we use these models to calculate the battery system response, estimate the internal states and parameters, and develop advanced BMSs. Systems design relies on analysis and simulation to estimate and optimize performance. The models developed in this book provide systems engineers with the tools to integrate batteries with the rest of the mechatronic system. State of charge (SOC) and state of health (SOH) estimators developed in Chapter 7 provide real-time measurements of the energy stored in the batteries and the total capacity of the batteries, respectively. Rather than using heuristic or empirical approaches, we can use the models developed to hard-wire the battery dynamics in the estimators developed, improving the accuracy and robustness of SOC and SOH estimation. Finally, the model-based BMSs discussed in Chapter 8 promise to deliver safe, efficient, and cost-effective battery systems for a variety of energy storage applications.

2

Electrochemistry

In this chapter we discuss the electrochemistry of three leading battery types: Pb–acid, Ni–MH, and Li-ion. For each battery type, the anode and cathode reactions are discussed, potential side reactions are introduced, and aging mechanisms are described. Finally, the performance of the battery chemistries is compared, including energy and power mass and volume densities, cost, and cycle life.

2.1 Lead–Acid

Pb–acid batteries are a relatively old technology that maintain 40–45% of the battery market, mainly due to their extensive use as starting, lighting, and ignition (SLI) batteries in automobiles, trucks, and buses [2]. They are also attractive for HEV and energy storage applications owing to their relatively high round-trip efficiencies of 75–80%. VRLA batteries are modern Pb–acid designs that immobilize the electrolyte using either highly porous and absorbent mats or a fumed silica gelling agent. Figure 2.1 shows an Enersys VRLA battery that uses adsorbed glass mat (AGM) plate separators to immobilize the electrolyte and allow ionic but not electrical conduction.

Figure 2.2 shows a schematic diagram of a VRLA battery cell composed of a positive electrode and a negative electrode with a separator in between that acts as an electronic insulator. All three components are porous and wholly or partially filled with an electrolyte (either liquid or solid). The electrolyte is an electronic insulator but a good conductor of the ionic species inside the cell. A gas phase may also be present in the cell if it is overcharged or overdischarged. The formation of gas phase is a side reaction that is not desirable and can lead to unsafe conditions and/or reduced battery life.

Reversible electrochemical reactions at the two electrodes allow the battery to be charged and discharged. The positive electrode is coated with lead oxide (PbO_2) and the negative electrode is made from Pb. Sulfuric acid (H_2SO_4) diluted with water (H_2O) acts as the electrolyte. The sulfuric acid dissociates into positive hydrogen ions (H^+) and negative ions (HSO_4^-) in water. At the negative electrode, Pb reacts with an HSO_4^- ion to produce lead sulfate ($PbSO_4$), an

Battery Systems Engineering, First Edition. Christopher D. Rahn and Chao-Yang Wang.
© 2013 John Wiley & Sons, Ltd. Published 2013 by John Wiley & Sons, Ltd.

Figure 2.1   Enersys VRLA group 31 battery (reproduced by permission of EnerSys)

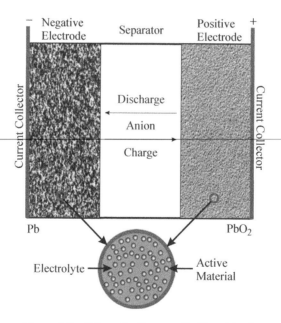

Figure 2.2   Schematic diagram of a Pb–acid cell

H^+ ion, and two electrons (e^-) during discharge. This is written in a chemical equation as follows:

$$Pb + HSO_4^- \underset{\text{charge}}{\overset{\text{discharge}}{\rightleftharpoons}} PbSO_4 + H^+ + 2e^-, \tag{2.1}$$

where the discharge reaction is indicated by the top, rightward-pointing arrow. The two electrons travel through the discharge circuit, providing power. The H^+ ions diffuse through the separator to the positive electrode. The charge reaction at the negative electrode simply reverses this process, recreating the Pb from $PbSO_4$.

At the positive electrode under discharge, PbO_2 reacts with one HSO_4^- ion, three H^+ ions that diffuse from the negative electrode, and two electrons from the external circuit to form $PbSO_4$ and two H_2O molecules as follows:

$$PbO_2 + HSO_4^- + 3H^+ + 2e^- \underset{\text{charge}}{\overset{\text{discharge}}{\rightleftharpoons}} PbSO_4 + 2H_2O. \tag{2.2}$$

The charge reaction returns the $PbSO_4$ to PbO_2.

The value of the cell voltage for a given battery chemistry depends on the anode and cathode materials. The negative, Pb electrode produces -0.3 V relative to a standard hydrogen electrode (SHE). The positive, PbO_2 electrode produces 1.6 V relative to an SHE. Thus, the overall Pb–acid cell voltage theoretically is 1.9 V. A cell with higher than 1.9 V is considered to be overcharged and a cell with less than 1.9 V is undercharged.

Under certain operating conditions, the Pb–acid reactions in Equations (2.1) and (2.2) are supplemented with side reactions or other processes that may reduce the efficiency of the cell and/or cause long-term degradation. We are most concerned with processes that age the battery and reduce its life. Operating conditions that promote these processes are to be avoided. Pb–acid battery life is governed by the following processes [3,4]:

- **Corrosion.** The positive electrode is Pb covered with a thin layer of PbO_2. The PbO_2 layer takes part in the reaction of Equation (2.2). The underlying Pb, however, can be corroded by the H_2SO_4 solution to form lead oxides that increase the resistance in the positive electrode.
- **Gas generation.** During overcharge, hydrogen evolves at the negative electrode and oxygen at the positive electrode, generating pressure inside the battery. If the pressure builds high enough, the valve in VRLA batteries opens, releasing the gas generated. This constitutes a permanent water loss that can dry out the separator and increase the acid concentration.
- **Sulfation.** Discharging the cell creates $PbSO_4$ crystals at both the positive and the negative electrodes. Charging converts the crystals back to the respective active material. Sulfation occurs when some crystals remain after charging, reducing the battery capacity. This is most likely to occur at elevated temperatures in partially discharged (low-voltage) cells that are either sitting idle or operating at very low discharge rates for long periods of time. The negative electrode is more prone to sulfation than the positive electrode is.
- **Active-material degradation.** Corrosion, gas generation, and sulfation can degrade the active material in the positive and negative electrodes. Mechanical stress is induced by gas generation and by the change in volume associated with charge and discharge. The specific volumes of $PbSO_4$ and PbO_2 are 2.4 and 1.96 times that of Pb, respectively, so significant

stresses develop in both the negative and positive electrodes as the cell discharges. As the SOC drops, the stresses increase. The thin PbO_2 layer on the positive electrode is most sensitive to material degradation because the PbO_2 sheds or sludges off the underlying Pb with prolonged cycling, reducing the effective active material.

- **Separator metallization.** The acid concentration in a Pb–acid cell changes with SOC because the reactions in Equations (2.1) and (2.2) consume acid during discharge. At low SOC, the acid is dilute, exacerbating the formation of $PbSO_4$. The $PbSO_4$ precipitate can fill the separator pores and, during charge, may be converted to dendritic, metallic lead. This can cause a short circuit through the separator.

2.2 Nickel–Metal Hydride

Ni–MH batteries offer higher performance at higher cost than VRLA batteries. They have very good cycle life and capacity and rapid recharge capability. They have been heavily used in HEV applications, including the Toyota Prius [5]. One drawback of Ni–MH batteries, however, is that they self-discharge relatively quickly without an applied load.

Figure 2.3 shows a schematic diagram of an Ni–MH cell. The positive electrode contains nickel hydroxide as its principal active material and the negative electrode is mainly composed of hydrogen-absorbing nickel alloys. The cell has an electrically insulating separator, an alkaline electrolyte (e.g., a solution of potassium hydroxide, KOH), and a vented metal case. With cylindrical Ni–MH batteries, the positive and negative electrodes are separated by the

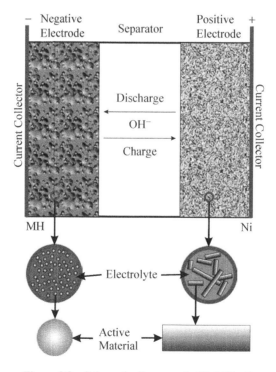

Figure 2.3 Schematic diagram of a Ni–MH cell

separator, wound into a coil, and inserted into the case. They can also be fabricated in prismatic form factors by stacking and electrically interconnecting multiple cells.

During discharge, nickel oxyhydroxide (NiOOH) is reduced to nickel hydroxide (Ni(OH)$_2$) in the positive electrode:

$$NiOOH + H_2O + e^- \underset{charge}{\overset{discharge}{\rightleftharpoons}} Ni(OH)_2 + OH^-, \tag{2.3}$$

where the hydroxide ions (OH$^-$) and electrons travel through the separator and external circuit, respectively. In the negative electrode, metal hydride (MH) is oxidized to the metal alloy (M):

$$MH + OH^- \underset{charge}{\overset{discharge}{\rightleftharpoons}} M + H_2O + e^-. \tag{2.4}$$

The negative and positive electrodes produce -0.83 V and 0.52 V versus SHE, respectively, yielding a theoretical cell voltage of 1.35 V.

Several side reactions occur in Ni–MH batteries that can be detrimental to their life and produce gases that can build up pressure inside the cell. During overcharge, the positive Ni electrode evolves oxygen:

$$4OH^- \underset{overcharge}{\longrightarrow} O_2 + 2H_2O + 4e^-. \tag{2.5}$$

The oxygen gas diffuses through the separator to the negative electrode, where it oxidizes MH to form water:

$$O_2 + 2H_2O + 4e^- \underset{overcharge}{\longrightarrow} 4OH^-. \tag{2.6}$$

The net effect of reactions (2.5) and (2.6) is that oxygen gas does not build up pressure. In extreme overcharge situations, however, hydrogen does form at the negative electrode in a potentially dangerous reaction:

$$2H_2O + 2e^- \underset{overcharge}{\longrightarrow} H_2 + 2OH^-. \tag{2.7}$$

Reaction (2.7) is typically not reversible, but the hydrogen gas can be absorbed in the active material. The relatively fast (on the order of days) self-discharge of Ni–MH cells, for example, is due to the reaction of hydrogen gas with NiOOH. Finally, the negative electrode can be corroded by water in the electrolyte in the side reaction

$$2M + H_2O \longrightarrow MH + MOH. \tag{2.8}$$

The gradual reduction in capacity that eventually limits the cycle life of Ni–MH batteries is governed by side reactions in the negative electrode [6]:

- **Hydrolysis.** Reaction (2.7) consumes water and produces hydrogen. The water loss causes cell dry-out, increasing the cell's internal resistance and the electrolyte concentration.

- **Corrosion.** The corrosion reaction (2.8) accelerates water loss by changing the balance between the positive and negative electrodes. The negative electrode is typically designed to have more active material so the capacity is limited by the positive electrode. During a full charge, oxygen evolves at the positive electrode according to reaction (2.5) and is consumed at the negative electrode by reaction (2.6). Reaction (2.7) is suppressed under these conditions unless high charge rates and/or low temperatures are present. At low- to medium-current operation, the small amount of hydrogen generated can be absorbed in the active material. Corrosion reduces the active material in the negative electrode, shifting the capacity balance toward the positive electrode and reducing the material available for hydrogen absorption. When the capacity of the positive electrode exceeds that of the negative electrode, hydrogen evolves at the negative electrode instead of oxygen at the positive electrode. The excess hydrogen cannot be absorbed in the active material and the cell pressure increases, potentially leading to venting and permanent water loss. High-temperature operation can increase water loss and accelerate corrosion.
- **Decrepitation.** Ni–MH cells also age through decrepitation of the active materials by induced stresses. The insertion and deinsertion of hydrogen in these materials causes lattice expansions and contractions that induce stresses. The amount of this hydrogen intercalation or de-intercalation depends on the depth of discharge (DOD). Higher DODs result in higher stresses and accelerated deterioration or decrepitation of the active materials.

2.3 Lithium-Ion

Li-ion batteries are commanding a greater market share owing to their high energy density, which makes them attractive for applications where weight or volume are important (e.g., HEVs). They have a long cycle life (>500 cycles) and low self-discharge rate ($<10\%$ per month). High initial cost has limited their use in price-sensitive applications, but new chemistries and economies of scale promise to reduce the cost of Li-ion batteries in the future.

Figure 2.4 shows a schematic diagram of an Li-ion cell. A lithium metal oxide ($LiMO_2$), where M stands for a metal such as Co, and lithiated carbon (Li_xC) are the active materials in the positive and negative electrodes, respectively. The metal in the positive electrode is a transition metal, typically Co. The active materials are bonded to metal-foil current collectors at both ends of the cell and electrically isolated by a microporous polymer separator film or gel-polymer. Liquid or gel-polymer electrolytes enable lithium ions (Li^+) to diffuse between the positive and negative electrodes. The lithium ions insert into or deinsert from the active materials via an intercalation process.

In the positive electrode during charge, the active material is oxidized and lithium ions are de-intercalated as follows:

$$Li_{1-x}CoO_2 + xLi^+ + xe^- \underset{charge}{\overset{discharge}{\rightleftharpoons}} LiCoO_2. \tag{2.9}$$

In the negative electrode during charge, the active material is reduced and lithium ions that migrate from the positive electrode and through the electrolyte and separator are intercalated in the reaction

$$Li_xC \underset{charge}{\overset{discharge}{\rightleftharpoons}} C + xLi^+ + xe^-. \tag{2.10}$$

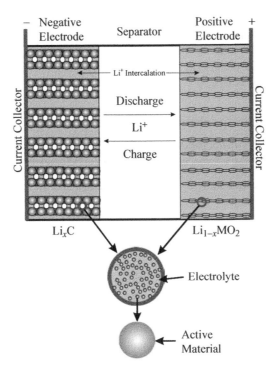

Figure 2.4   Schematic diagram of a Li-ion cell

Reactions (2.9) and (2.10) reverse for discharge. These reactions produce a theoretical cell voltage of 4.1 V, much higher than either the Ni–MH or Pb–acid cells.

The power and energy capacity of Li-ion batteries fade with cycling. Power fade is primarily due to an increase in internal resistance or impedance. Internal resistance causes ohmic losses that waste energy, produce heat, and accelerate aging. In addition to ohmic energy losses, Li-ion batteries lose capacity over time due to degradation of the positive and negative electrodes and the electrolyte. The degradation mechanisms are complex, coupled, and dependent on cell chemistry, design, and manufacturer [7].

In the negative electrode, the dominant aging mechanisms [8] are:

- **Solid–electrolyte interface (SEI) layer growth.** The SEI layer grows on the negative electrode, leading to an impedance rise. The SEI layer forms at the beginning of cycling and grows during cycling and storage, especially at higher temperatures. The SEI layer entrains lithium.
- **Lithium corrosion.** Lithium in the active carbon material of the negative electrode can corrode over time, causing capacity fade due to irreversible loss of mobile lithium.
- **Contact loss.** The SEI layer disconnects from the negative electrode, leading to contact loss and increased cell impedance.
- **Lithium plating.** Lithium metal can plate on the negative electrode at low temperatures, high charge rates, and low cell voltages, causing irreversible loss of cyclable lithium.

Recent studies have shown that impedance rise and capacity fade during cycling are primarily caused by the positive electrode [9]. The discharge capacity may be limited by a decrease in active lithium intercalation sites in the oxide particles. A passivation layer also forms on the positive electrode and it thickens and changes properties during cycling, causing cell impedance rise and power fade.

2.4 Performance Comparison

2.4.1 Energy Density and Specific Energy

Table 2.1 compares Pb–acid, Ni–MH, and Li-ion batteries in several key categories. The theoretical voltage is determined by the electrode materials and the practical voltage is what can be achieved in a real battery. The actual values are essentially the same as the theoretical values for Pb–acid and Li-ion batteries at roughly 2 V and 4 V, respectively. The practical voltage of Ni–MH batteries is around 10% lower than the theoretical value. The specific energy is the energy storage capacity of the battery in watt-hours (Wh) divided by the mass of the battery in kilograms (kg). The theoretical capacity in ampere-hours/gram (Ah/g) is based on the equivalent weight of the active materials participating in the electrochemical reaction. Multiplying the theoretical capacity and voltage gives the theoretical specific energy in Wh/kg. Pb–acid has the lowest capacity at 166 Wh/kg and Ni–MH is around 250 Wh/kg. The Li-ion battery theoretically yields a much larger 410 Wh/kg specific energy. In practice, however, none of the chemistries lives up to its theoretical potential. Li-ion is still the highest at 150 Wh/kg, almost a third of the theoretical value. Ni–MH batteries demonstrate 75 Wh/kg in practice, also around a third of theoretical. Pb–acid cells only perform at 14% of their theoretical specific energy, or 35 Wh/kg. The high energy density of Li-ion batteries at 400 Wh/L is 1.7 and 5.7 times higher than Ni–MH and Pb–acid, respectively. It is clear that Li-ion batteries have an advantage in weight- and volume-sensitive applications like HEVs.

The energy storage efficiency is also an important metric for batteries. Two efficiency metrics are used: coulometric efficiency and energy efficiency. Coulometric efficiency

$$f = \frac{\int_{discharge} I \, dt}{\int_{charge} I \, dt},$$ (2.11)

Table 2.1 Comparison of Pb–acid, Ni–MH, and Li-ion performance

	Pb–acid	Ni–MH	Li-ion
Theoretical			
Voltage (V)	1.93	1.35	4.1
Specific energy (Wh/kg)	166	240	410
Practical			
Specific energy (Wh/kg)	35	75	150
Energy density (Wh/L)	70	240	400
Coulometric efficiency	0.80	0.65–0.70	>0.85
Energy efficiency	0.65–0.70	0.55–0.65	~0.80
Specific power, 80% DOD (W/kg)	220	150	350
Power density (W/L)	450	>300	>800

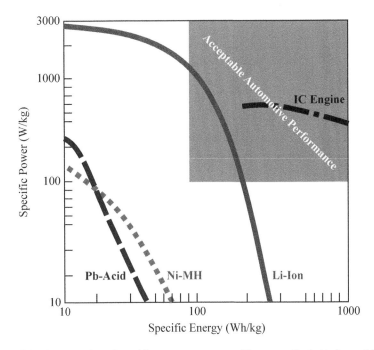

Figure 2.5   Ragone plot of specific power versus specific energy for batteries and ICEs

where $I(t)$ is the battery current and t is time. Energy efficiency

$$\eta = \frac{\int_{\text{discharge}} IV \, dt}{\int_{\text{charge}} IV \, dt}, \tag{2.12}$$

where $V(t)$ is the battery voltage. Table 2.1 shows that Li-ion is the most efficient chemistry followed by Pb–acid and Ni–MH.

Figure 2.5 is a Ragone plot of the specific power versus specific energy of the three battery chemistries and ICEs. The region for acceptable automotive performance is highlighted. ICEs provide the desired specific power and energy for this application. Batteries that can operate in this region could be used for pure EV applications. Li-ion cells meet these criteria, making them a viable option for EVs. Pb–acid and Ni–MH batteries, however, fall short of the power and energy requirements of an EV.

2.4.2   Charge and Discharge

The dynamic performance of a battery in charge and discharge operation regulates the speed at which current can be put into and taken from storage. The terminal voltage does not remain constant during charge and discharge, but rises and falls during steady charging and discharging, respectively. After charging or discharging has stopped, the transient voltage response settles out in seconds to minutes. For sufficiently long charge, the battery voltage saturates at a maximum value. In these overcharge situations, most of the input energy goes to

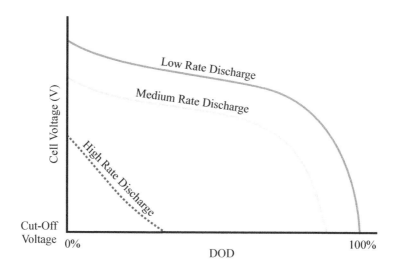

Figure 2.6   Example voltage curves for different discharge rates

heating losses or side reactions that can be harmful to the battery. Similarly, undercharge occurs when the battery voltage falls below the end or cut-off voltage due to excessive discharge and can cause damage to the battery.

The working range of a battery is determined by the SOC, defined as the percentage of maximum possible charge that is present inside a rechargable battery [10]. A fully charged battery is at 100% SOC. The working SOC range depends on the application. HEVs with Ni–MH batteries, for example, typically operate from 30% to 70% SOC. In this SOC range the battery has very high coulometric efficiency. The DOD (DOD = 100% − SOC) is another way to quantify stored charge.

The charge and discharge dynamics of batteries can be characterized by measurements of voltage under constant charge and discharge current inputs. The rate of charge or discharge is measured relative to the battery capacity C. For example, a $0.1C$ discharge rate for a 5 Ah battery is 0.5 A or a $2C$ discharge rate for a 10 Ah battery is 20 A. Figure 2.6 shows representative plots of a battery being discharged at low, medium, and high rates. As the current rate is constant, one can plot the voltage response versus time or DOD. The low rate curve approximates the equilibrium cell (or open-circuit) potential. The best open-circuit potential curve is flat over a broad range of DOD so that the voltage remains essentially constant during discharge, simplifying the design and reducing the cost of the associated voltage-regulation circuitry. The medium rate discharge curve is shifted downward due to ohmic losses over the entire DOD range and charge transfer kinetic losses and mass transport limitations at low and high DOD, respectively. The high rate discharge case demonstrates that the voltage drops quickly so that only a fraction of the capacity can be utilized at high discharge rates.

Specific power (W/kg) and power density (W/L) are good summary statistics for battery discharge performance. Table 2.1 shows that Li-ion has the highest specific power and power density, Pb–acid comes in second, and Ni–MH is the least powerful. The discharge profile of Ni–MH is the flattest, however, followed by Pb–acid and Li-ion, which have more sloped discharge profiles.

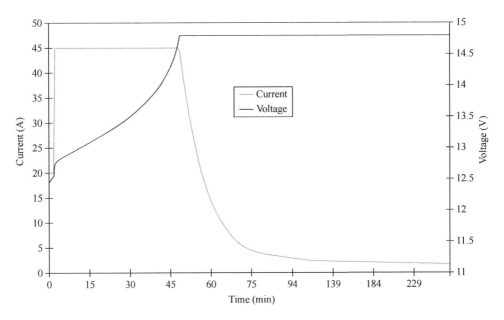

Figure 2.7   Constant-current and constant-voltage charge curves

Charging batteries from wall power replaces the energy expended during discharge. The charger can either control current or voltage during charging, not both. The relationship between current and voltage is dictated by the battery impedance. Charging typically consists of periods of constant current (CC) and/or constant voltage (CV) charging. Figure 2.7 shows an example CC–CV charging profile for a Pb–acid battery. Initially, when the battery has a low SOC, a CC charge is applied to bring the voltage up to the CV level. If the CV charge is applied during this initial phase, the current will be too high and side reactions or excessive temperature rise will result. Once the desired voltage level has been achieved, the charger switches to CV mode and the current decreases as the battery reaches 100% SOC. Leaving the battery on CV mode for long periods of time is called a float charge. Similarly, a trickle charge maintains CC mode at a low ($C/100$) rate. More sophisticated chargers use multiple CC and CV steps of different current and voltage levels, tapered current–time trajectories, and additional sensors (e.g., temperature and pressure) to minimize the potential damage to the battery and maximize life, safety, and efficiency. Fast charge controllers enable high current charging but use feedback to cut off the charging current before an excessive rise in gassing, pressure, or temperature occurs.

Pb–acid, Ni–MH, and Li-ion batteries can all be CC charged. Pb–acid and Ni–MH batteries can be slightly overcharged without dire consequences, but gas generation and overheating are possible with excessive overcharging. Li-ion batteries, however, are irreversibly degraded and may vent during overcharge. Pb–acid and Li-ion batteries can be CV charged, but this is not recommended for Ni–MH batteries. A CC approach with stepped or tapered current is recommended for Ni–MH batteries and the CC–CV method shown in Figure 2.7 is recommended for Pb–acid and Li-ion batteries [2].

The charge acceptance of a battery dictates how fast it can be charged. Li-ion, Ni–MH, and Pb–acid are recommended to be charged at $C/3$. In fast charge mode, Pb–acid batteries can be charged at up to $4C$. Ni–MH batteries can be fast charged at as high as $1C$ with a controller that terminates charge when the voltage and/or temperature rates exceed thresholds. Li-ion batteries accept $2C$ as a high rate of charge, but excessively low and high cell voltages should be avoided.

2.4.3   Cycle Life

The cycle life of a battery depends on the chemistry, discharge–charge cycle, temperature, prior history (e.g., storage), and manufacturer. The most accurate and reliable way to determine cycle life is to test several batteries from the same batch to determine fresh cell capacity. Then, the batteries are tested on a cycling machine that repeats a prescribed current trajectory (e.g., DST or SFUDS) that represents a typical cycle in the proposed application. At uniform intervals, one of the batteries is removed from cycling and tested for capacity. In this way, a plot of capacity versus number of cycles can be obtained.

All three chemistries studied in this book have good cycle life. In general, batteries have longer life for lower DOD cycles. Li-ion cells have the longest cycle life. At 100% DOD they typically last 3000 cycles at low discharge/charge rates and room temperature. At 20–40% DOD they can last up to 20 000 cycles. Ni–MH batteries last around 500 cycles at 80% DOD at $0.2C$ charge/discharge and room temperature. For Pb–acid batteries, 200 cycles are typical at 100% DOD. Up to 1500 cycles can be attained at 25% DOD. The end of life is characterized by a drop in capacity to 50–80% of the initial capacity, depending on the chemistry and application.

2.4.4   Temperature Operating Range

Batteries perform poorly at extremely low and high temperatures. At low temperature, ionic diffusion and migration can be hindered and damaging side reactions (e.g., lithium plating) can occur. High temperatures favor other side reactions, such as corrosion and gas generation. The discharge operating range of a battery typically has a lower lower limit and higher upper limit than the charge operating range. For Pb–acid batteries, charge and discharge temperatures should be limited to an operating range of -40–$60°$C. Li-ion batteries have an operating range of -20–$60°$C. Ni–MH have the narrowest operating range of -20–$45°$C.

3

Governing Equations

To gain insight into the physical phenomena that govern the response of batteries, it is essential for the battery systems engineer to understand the origin and derivation of the governing equations. This provides a solid foundation for the model development, simplification, and model-based estimation and control that are the topics of subsequent chapters. It also helps develop a physical feel for the underlying processes, parameters, and assumptions in the derived battery models. This chapter provides an overview suitable for battery systems engineers. Further information on this topic can be found in [11].

In this chapter, the governing equations of battery dynamics are derived from fundamental conservation laws. The thermodynamic properties that generate electrochemical potentials are presented. The electrokinetics that relate current flow to the battery overpotential are introduced. The conservation laws that govern ion (species) and charge transport are derived. The various potentials developed in the cell are related to the overall cell voltage. The cell heat generation and dissipation are modeled. Side-reaction and aging models are discussed.

3.1 Thermodynamics and Faraday's Law

The voltage that a cell produces at zero current is called the equilibrium or open-circuit potential U. This voltage depends on the thermodynamics of the anode and cathode, where the reactions in the cell occur. During discharge these reactions produce and consume electrons at the negative and positive electrodes, respectively. This causes current to flow from the positive electrode to the negative electrode. The current produced by a cell is related to the consumption of active material by Faraday's law.

The reactions at the anode and cathode can be expressed as

$$\sum_k s_k \mathcal{M}_k^{z_k} \rightleftharpoons n e^-, \tag{3.1}$$

where, for the species k, $\mathcal{M}_k$ is the symbol for the chemical formula, s_k is the stoichiometry, and z_k is the charge number. In general, the reactions at the anode and cathode have different

Battery Systems Engineering, First Edition. Christopher D. Rahn and Chao-Yang Wang.
© 2013 John Wiley & Sons, Ltd. Published 2013 by John Wiley & Sons, Ltd.

$\mathcal{M}_k$, s_k, and z_k but produce the same number of electrons n. Conservation of charge requires that

$$\sum_k s_k z_k = -n. \tag{3.2}$$

Only the charged species matter in Equation (3.2) because $z_k = 0$ for the uncharged species. For Li-ion and Ni–MH batteries there is only one charged species (unary electrolyte), so $sz = -n$. Pb–acid cells have a binary electrolyte with two charged species, so $s_+ z_+ + s_- z_- = -n$. The reactions in Equation (3.1) involve the production and consumption of electrons. Closing the circuit through a load will result in the flow of current from the positive electrode to the negative electrode.

By Faraday's first law, the current $I(t)$ produced by the anode and cathode reactions is related to the consumption or production of active material as follows:

$$\frac{dm_k}{dt} = \begin{cases} -\frac{s_k M_k}{nF} I & \text{in the negative electrode} \\[2mm] \frac{s_k M_k}{nF} I & \text{in the positive electrode} \end{cases} \tag{3.3}$$

where m_k (g) is the mass of active material, M_k (g/mol) is the molecular weight of species k, and $F = 96\,485$ C/mol is the Faraday constant. The current flows from the positive electrode to the negative electrode (opposite of electron flow), so positive current signifies the loss of electrons from the negative electrode and gain of electrons on the positive electrode.

For example, the reaction at the negative electrode of a Pb–acid cell under discharge is

$$Pb + HSO_4^- \rightarrow PbSO_4 + H^+ + 2e^-, \tag{3.4}$$

where: $\mathcal{M}_1 = Pb$, $s_1 = 1$, $z_1 = 0$; $\mathcal{M}_2 = HSO_4^-$, $s_2 = 1$, $z_2 = -1$; $\mathcal{M}_3 = PbSO_4$, $s_3 = -1$, $z_3 = 0$; $\mathcal{M}_4 = H^+$, $s_4 = -1$, $z_4 = 1$; and $n = 2$. The mass change of Pb for a 1 Ah cell is

$$\Delta m_{Pb} = -\frac{s_1 M_1}{nF} It = -\frac{207.2 \text{ g/mol}}{2(96\,485 \text{ C/mol})}(1 \text{ C/s})(3600 \text{ s}) = -3.87 \text{ g}. \tag{3.5}$$

The reaction at the positive electrode during discharge is

$$PbO_2 + HSO_4^- + 3H^+ + 2e^- \rightarrow PbSO_4 + 2H_2O. \tag{3.6}$$

where: $\mathcal{M}_1 = PbO_2, s_1 = -1, z_1 = 0$; $\mathcal{M}_2 = HSO_4^-, s_2 = -1, z_2 = -1$; $\mathcal{M}_3 = PbSO_4, s_3 = 1$; $z_3 = 0$, $\mathcal{M}_4 = H^+$, $s_4 = -3$, $z_4 = 1$; $\mathcal{M}_5 = H_2O$, $s_5 = 2$, $z_5 = 0$; and $n = 2$. The mass change of PbO_2 for a 1 Ah cell is

$$\Delta m_{PbO_2} = -\frac{239.2 \text{ g/mol}}{2(96\,485 \text{ C/mol})}(1 \text{ C/s})(3600 \text{ s}) = -4.46 \text{ g}. \tag{3.7}$$

The HSO_4^- and H^+ ions come from dissolved H_2SO_4. The acid is consumed in both the positive and negative electrodes during discharge. The H^+ ions produced in the negative electrode migrate to the positive electrode. Acid is converted to $PbSO_4$ in both electrodes at the same rate; so, for a 1 Ah cell,

$$\Delta m_{H_2SO_4} = -\frac{98.1 \text{ g/mol}}{2(96\,485 \text{ C/mol})}(1 \text{ C/s})(3600 \text{ s}) = -1.83 \text{ g} \qquad (3.8)$$

of H_2SO_4 is consumed at each electrode.

This example shows the need to balance the various species in a cell to maximize energy storage. Acid, Pb, and PbO_2 are all consumed during discharge but at different rates. For maximum energy storage, all three active materials should be expended at the same time. This means that the materials must be available in the proper mass ratios ($m_{PbO_2} = 1.15 m_{Pb} = 1.21 m_{H_2SO_4}$). The reactions occur at the surface of the solid phase so the anode and cathode are composed of porous material to maximize surface area. The mass and porosity of the positive and negative plates are carefully balanced with the acid mass in a high performance Pb–acid cell.

One can use Faraday's law to calculate the theoretical specific energy of a given cell chemistry. Consider, for example, the Pb–acid cell with the combined anode and cathode reaction:

$$Pb + PbO_2 + 2\,H_2SO_4 \underset{\text{charge}}{\overset{\text{discharge}}{\rightleftharpoons}} 2\,PbSO_4 + 2\,H_2O. \qquad (3.9)$$

This produces a theoretical potential of 1.93 V. Two electrons are generated during the reaction generating $2F = 2$ mol $\times 96\,485$ C/mol $= 192\,970$ As $= 53.61$ Ah. Table 3.1 calculates the theoretical mass of the reactants to be 646 g. The theoretical specific energy of a Pb–acid battery cell equals

$$53.61 \text{ Ah} \times 1.93 \text{ V}/642.6 \text{ g} = 166 \text{ Wh/kg} = 597.6 \text{ kJ/kg}. \qquad (3.10)$$

The thermodynamics of the cell are represented by the Gibbs–Helmholtz equation:

$$\Delta H = \Delta G + T \Delta S, \qquad (3.11)$$

where ΔH (J) is the enthalpy, ΔG (J) is the Gibbs free energy, T (K) is absolute temperature, and ΔS (J/K) is the entropy of the reaction. The Gibbs free energy represents the useful work that can be done by the cell. The enthalpy is the theoretically available energy from the reaction

Table 3.1 Theoretical mass of Pb–acid reactants

Reactants	Molar mass (g/mol)	Moles (mol)	Total mass (g)
Pb	207.2	1	207.2
PbO_2	239.2	1	239.2
H_2SO_4	98.1	2	196.2
Total			642.6

and $T\Delta S$ is the amount of heat generated (or wasted energy). The open-circuit potential is related to the Gibbs free energy as follows:

$$U = -\frac{\Delta G}{nF}. \tag{3.12}$$

The standard cell potentials U^{θ} have been tabulated for a wide variety of electrodes. For the Pb–acid, Ni–MH, and Li-ion batteries that are the focus of this book, the open-circuit potentials are provided in Table 2.1.

The open-circuit potential also depends on the concentration of the various species involved in the reaction:

$$U = U^{\theta} + \Delta U(c, T), \tag{3.13}$$

where U^{θ} is the constant part of the open-circuit potential, tabulated in Table 2.1 at a specific temperature and concentration, and ΔU incorporates temperature and concentration effects. One version of Equation (3.13) is the Nernst equation that uses logarithms of the species concentrations in ΔU. Equation (3.13) linearizes to

$$U \approx \bar{U} + \frac{\partial \Delta U}{\partial c}(c - \bar{c}), \tag{3.14}$$

where we linearize at a constant concentration $c(x, t) = \bar{c}$ and $\bar{U} = U^{\theta} + \Delta U(c_0, T)$.

The reversible reaction heat of the cell Q (J) or Peltier effect can be calculated from

$$Q = T\Delta S = nFT\frac{\partial U}{\partial T} \tag{3.15}$$

and the rate of heat generation for an isothermal cell is

$$\dot{Q} = I\left(V - U + T\frac{\partial U}{\partial T}\right). \tag{3.16}$$

The Peltier effect is reversible, so charge and discharge can cause cooling as well as heating. The ohmic losses in a cell, especially at high current rates, often outweigh the Peltier effect, resulting in cells that heat up during both charge and discharge.

3.2 Electrode Kinetics

Pb–acid, Ni–MH, and Li-ion batteries have porous, solid-phase electrodes that are saturated with liquid-/gel-phase electrolyte that transports ions from one electrode to the other. The electrons involved in the electrode reactions must pass through the electrode–electrolyte interface. This interface resists the flow of electrons, creating an overpotential that must be overcome to allow charge transfer. The overpotential is defined as

$$\eta = \phi_{s} - \phi_{e} - U, \tag{3.17}$$

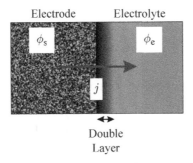

Figure 3.1   Electrode kinetics at the SEI

where ϕ_s (V) and ϕ_e (V) are the electrode and electrolyte potentials, respectively, as shown in Figure 3.1. To overcome the energy barrier associated with the surface reaction, overpotential is produced. For anodic or oxidation reactions, current flows from the electrode to the electrolyte and $\eta > 0$. For cathodic or reduction reactions, current flows in the opposite direction and the overpotential is negative.

3.2.1   The Butler–Volmer Equation

For small current density charge or discharge (j, mA/cm^2), the electrode reactions in Equation (3.1) generate overpotential as follows:

$$\eta = R_{ct}j, \tag{3.18}$$

where R_{ct} (Ω cm^2) is the charge transfer resistance. For large $|j|$, the overpotential is proportional to the log of the current, where the proportionality constant is referred to as the Tafel slope. The Butler–Volmer equation combines these two characteristics of electrode kinetics as follows:

$$j = i_0 \left[\exp\left(\frac{\alpha_a F}{RT}\eta \right) - \exp\left(\frac{-\alpha_c F}{RT}\eta \right) \right], \tag{3.19}$$

where i_0 (mA/cm^2) is the exchange current density, α_a and α_c are the apparent exchange coefficients, and R (8.3143 J/(mol K)) is the universal gas constant. The exchange current density can vary over a wide range (1–10^{-7} mA/cm^2) depending on the concentrations of the reactants and products, temperature, and the nature of the SEI. The apparent exchange coefficients typically range from 0.2 to 2. They are related to the number of electrons involved in the reaction

$$\alpha_a + \alpha_c = n \tag{3.20}$$

and $\alpha_a \approx \alpha_c \approx n/2$.

Linearization of Equation (3.19) around $\eta = 0$ yields

$$R_{\mathrm{ct}} = \frac{RT}{nF i_0},\tag{3.21}$$

using Equation (3.20). For large, positive overpotential ($\eta \to \infty$),

$$j \to j_{\mathrm{a}} = i_0 \exp\left(\frac{\alpha_{\mathrm{a}} F}{RT}\eta\right)\tag{3.22}$$

and the anodic reaction dominates. For $\eta \to -\infty$, the cathodic reaction dominates and

$$j \to j_{\mathrm{c}} = -i_0 \exp\left(-\frac{\alpha_{\mathrm{c}} F}{RT}\eta\right).\tag{3.23}$$

Taking the logarithm of Equations (3.22) and (3.23) yields

$$\eta_{\mathrm{a}} = \frac{RT \ln(10)}{\alpha_{\mathrm{a}} F} \log\left(\frac{j_{\mathrm{a}}}{i_0}\right),\tag{3.24a}$$

$$\eta_{\mathrm{c}} = -\frac{RT \ln(10)}{\alpha_{\mathrm{c}} F} \log\left(\frac{-j_{\mathrm{c}}}{i_0}\right),\tag{3.24b}$$

where the Tafel slopes $2.303 RT/\alpha F$ depend on the apparent transfer coefficient.

Figure 3.2 shows the Tafel approximations in Equations (3.24a) and (3.24b) and the Butler–Volmer equation (3.19). The Butler–Volmer equation converges to the anodic (top) and cathodic (bottom) Tafel approximations as $\eta \to \pm\infty$. For small η, the Butler–Volmer equation is approximately linear with a slope of $1/R_{\mathrm{ct}}$. The linear range is approximately ± 50 mV for this case.

3.2.2 Double-Layer Capacitance

A thin (on the order of nanometers) double layer often forms at the SEI that provides capacitance in parallel with the charge transfer resistance on the order of 30 μF/cm^2. Double-layer capacitance acts in parallel to the Butler–Volmer equation (3.19), producing a current

$$i_{\mathrm{dl}} = C_{\mathrm{dl}} \frac{\partial \eta}{\partial t},\tag{3.25}$$

where C_{dl} (F/cm^2) is the double-layer capacitance. Inclusion of this effect increases the model order and typically effects only the high-frequency dynamics. The battery system engineer must decide whether the increased complexity is warranted for a given application.

3.3   Solid Phase of Porous Electrodes

Electrodes in battery cells can be planar or porous. Planar electrodes consist of a uniform foil or plate that is coated or entirely formed of active material. Porous electrodes provide greater

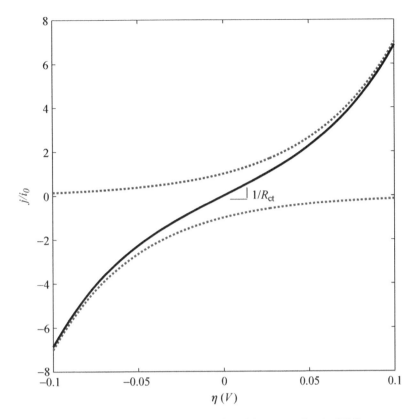

Figure 3.2 Tafel approximations (dotted) and Butler–Volmer equation (solid) for $\alpha_a = \alpha_c = 0.5$

interfacial area so more active material is exposed to the electrolyte and can be consumed in the reaction.

Most batteries have porous electrodes that are saturated with electrolyte, as shown in Figure 3.3. The large interfacial area of porous electrodes can accelerate the electrochemical reactions, allowing faster charge and discharge. Porous electrodes are compact and, therefore, have smaller ohmic resistances because the current flows through a shorter length of resistive material. This reduces the losses and potential for side reactions in battery cells.

Figure 3.3 shows that porous electrodes provide multiple pathways for current to flow. The porous solid-phase material in the negative electrode extends from the negative terminal to the separator. Similarly, the saturating electrolyte liquid/gel phase conducts current over the same domain, through the separator, and into the positive electrode. The positive electrode extends from the separator to the positive terminal. At the interface between the solid and electrolyte phases there is an open-circuit potential generated that drives current flow. The charge transfer resistance associated with Butler–Volmer kinetics also acts at this interface.

The solid and electrolyte phases have conductivity σ and κ, respectively. The porosity of the electrode means that current is distributed between the solid and electrolyte phases, conceptually shown as the series resistors at the top and bottom of Figure 3.3. The cell voltage

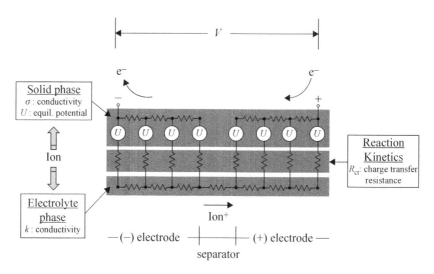

Figure 3.3 Porous electrode and equivalent circuit

equals the difference in solid phase potential between the left and right current collectors at the two ends of the cells where the battery terminals are connected.

At the microscale, porous electrodes have complex geometries that are hard to characterize and analyze. To overcome this difficulty, volume-averaged quantities are used to produce a macroscopic model that captures the large-scale phenomena of interest. Porous electrode models are based on conservation of species and conservation of charge in the solid and electrolyte phases.

3.3.1   Intercalate Species Transport

Electrons flow in the conductive matrix material of the electrodes in all cells, carrying the charge/discharge current. Li-ion and Ni–MH cells employ intercalation compound electrodes involving reversible insertion/extraction of active species into/from the host material during the discharge/charge process. Active material particles that make up the electrode in an Li-ion or Ni–MH cell can be approximately spherical and cylindrical, respectively. The conservation of active species within a particle is governed by Fick's law of diffusion,

$$\frac{\partial c_s}{\partial t} = D_s \left(\frac{\partial^2 c_s}{\partial r^2} + \frac{m}{r} \frac{\partial c_s}{\partial r} \right) \qquad \text{for } r \in (0, R_s), \qquad (3.26)$$

where $r \in (0, R_s)$ is the radial coordinate, $c_s(r, t)$ (mol/cm^3) is the concentration of lithium or protons in the host particle as a function of radial position r and time t, and D_s (cm^2/s) is the solid-phase diffusion coefficient. For the spherical Li-ion particles, $m = 2$. For the cylindrical Ni–MH particles, we may assume an infinite length (neglect the effect of diffusion through the ends of the cylinder) and use $m = 1$.

Fick's law of diffusion in Equation (3.26) arises from species conservation within the particle. The active species are distributed throughout the particle in a spherically or cylindrically symmetric arrangement. Equation (3.26) results from taking a differential radial element of the particle and performing a mass balance. Taking the limit of small element size produces the PDE (3.26). The boundary conditions are

$$\frac{\partial c_s}{\partial r}\bigg|_{r=0} = 0, \tag{3.27}$$

$$\frac{\partial c_s}{\partial r}\bigg|_{r=R_s} = \frac{-j}{D_s F}, \tag{3.28}$$

so the species flux in and out of the particle surface ($r = R_s$) with zero flux at the center of the particle ($r = 0$).

3.3.2 Conservation of Charge

Batteries are three-dimensional but we assume that the potential and concentration distributions in the solid and electrolyte are uniform across the two dimensions in the plane of the cell and focus only on the dimension x from the negative electrode to the positive electrode. Cells are designed to have uniform concentration in the y–z plane. The materials and geometry are uniform. The boundary conditions on the x–y and x–z faces of the cell enforce zero charge and concentration flux, minimizing gradients in the plane of the cell. The y–z faces of the cell are attached to the electrodes so that charge fluxes in and out of the cell, resulting in nonuniform potential and concentration distributions along the x axis. Electrolyte convection can cause three-dimensional flows in the cell, but these effects are typically negligible for VRLA, Ni–MH, and Li-ion batteries.

Charge concentration or electric potential in the solid phase follows Ohm's law:

$$\sigma^{\text{eff}} \frac{\partial^2 \phi_s}{\partial x^2} = a_s j, \tag{3.29}$$

where $\phi_s(x, t)$ (V) is the solid-phase potential, σ^{eff} ($1/(\Omega \text{ cm})$) is the effective conductivity, and a_s ($1/\text{cm}$) is the specific interfacial area. The specific interfacial area is a principal geometric parameter for characterizing electrode performance. It depends on the morphology of the electrode, including the porosity and grain size, and the cell SOC (e.g., for conversion reactions). The specific interfacial area is the surface area of the active material per unit volume of the electrode. Porous electrodes extend from a current collector to the separator. The resistance of the electrode is typically much smaller than that of the electrolyte. The electrode solid-phase conductivity is typically high (ranging from 10^{-1} to $10^4 S/\text{cm}$, where $S = 1/\Omega$) because the solid phase consists of low-resistance metallic compounds or carbon.

The porosity of the electrode leads to the definition of effective parameters such as σ^{eff}. The conductivity of the uniform (nonporous) solid phase is σ and the porosity is ε. The porosity varies in the range $0 < \varepsilon < 1$, with $\varepsilon = 0$ corresponding to a solid electrode. The conductivity of the solid phase depends on the composition of the electrode (e.g., active materials, conductive fillers, and binders) and how the electrode is fabricated. Porosity is defined as the volume of the pores in the electrode divided by the total volume of the electrode. If the pores are fully saturated with electrolyte, then the porosity is also the volume fraction of electrolyte in the electrode. The electrodes consist of solid phase and electrolyte, so the volume fraction of the solid phase is $1 - \varepsilon$. The effective conductivity of the electrode is smaller than the conductivity of the bulk material owing to porosity according to the Bruggeman relation:

$$\sigma^{\text{eff}} = \sigma(1 - \varepsilon)^{1.5}, \tag{3.30}$$

where the extra 0.5 in the exponent accounts for the effect of electrode tortuosity on the conductance. Here, we assume that the electrode is composed of only two phases, solid and electrolyte, so that $\varepsilon_e = 1 - \varepsilon_s = \varepsilon$. In some cases, it may be necessary to define separate electrolyte ε_e and solid ε_s phase porosities if there are active particles embedded in an inactive matrix or inactive fillers/binders in the electrode. For active spherical particles embedded in the electrode, $a_s = 3\varepsilon_s/R_s$, where R_s is particle radius. One can also think of σ^{eff} as the conductivity that one would measure across the electrode if the electrolyte were removed. The boundary conditions on the solid-phase electrode are related to the flux of charge in and out of the electrode. If a boundary of the electrode is connected to a current collector, then charge can flux according to

$$\sigma^{\text{eff}} \frac{\partial \phi_s}{\partial x} = \pm \frac{I}{A}, \tag{3.31}$$

where $I(t)$ (A) is the current and A (cm^2) is the electrode plate area. At an electrically insulated boundary, no current flows, so

$$\frac{\partial \phi_s}{\partial x} = 0. \tag{3.32}$$

Positive current flows from high potential to low potential. Figure 3.4 shows the relationships between current and the gradient of solid-phase potential for current collectors on the left and right side of a porous electrode with the x-axis going from left to right. In the left current collector case, for $\phi_s' = \partial \phi_s/\partial x > 0$ the potential increases with increasing x. Thus, the current flows to the left, in the direction of decreasing potential. Similarly, for the current collector on the right side of a porous electrode, the current flows to the left for $\phi_s' > 0$.

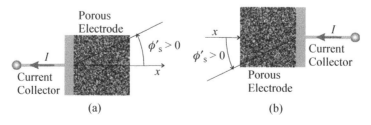

Figure 3.4   Solid-phase potential and current flow with current collectors at (a) left and (b) right

3.4   Electrolyte Phase of Porous Electrodes

The electrolyte transports ions from the negative electrode to the positive electrode. For Li-ion and Ni–MH batteries, only one ion (Li^+ and OH^-, respectively) participates in the reactions at the anode and cathode, so the electrolyte is unary. For Pb–acid cells, however, both the positive and negative ions participate in the charge and discharge reactions and migrate from one electrode to the other, resulting in a binary electrolyte. The transport response of a binary electrolyte depends on the properties of both ions.

The unary electrolyte of Ni–MH and Li-ion cells transports a single type of ion. The ions insert or deinsert in the active material depending on the direction of the current. During this intercalation process the ion enters or leaves the crystal lattice of the solid electrode and leaves or enters the electrolyte. The ions originate in one insertion electrode, travel through the electrolyte, and intercalate in the other insertion electrode. Electrons enter or leave the crystal to maintain electroneutrality and produce the desired current flow.

3.4.1   Ion Transport

Transport in the electrolyte phase of batteries involves the movement of charged species (ions) across the cell and the corresponding changes in ionic concentration distribution $c(x, t)$ that depend on the position in the cell x and time t. Ni–MH and Li-ion batteries have one kind of charged species that travels through the unary electrolyte. For Pb–acid batteries, however, each sulfuric acid molecule (H_2SO_4) dissociates into one H^+ positive ion and one HSO_4^- negative ion. For these binary electrolytes, electroneutrality requires that the molar concentration of the acid in mol/cm^3 is

$$c = \frac{c_+}{v_+} = \frac{c_-}{v_-},\tag{3.33}$$

where c_i is the concentration of the ion and v_i is the number of ions produced by the dissociation of one molecule of electrolyte. For Pb–acid cells, we have $v_+ = v_- = 1$, so the concentrations of H^+ and HSO_4^- ions equal the sulfuric acid concentration. Thus, for all three battery types there is one concentration $c(x, t)$ (mol/cm^3) that governs ion transport.

Conservation of the ion mass requires that the accumulation of ions equals the net input of ions plus the production of ions. In battery cells the reactions occur at the electrode surfaces, so there is no production of ions inside the electrolyte. For porous electrodes, however, the solid phase is distributed throughout the electrolyte, adding source terms to the governing equations similar to ion production. Thus, conservation of mass is expressed as

$$\varepsilon\frac{\partial c}{\partial t} = a_s J - \frac{\partial N}{\partial x},\tag{3.34}$$

with $N(x, t)$ ($mol/(cm^2\ s)$) being the flux density of the ions and $J(x, t)$ ($mol/(cm^2\ s)$) the flux density of ions from the solid phase.

At the cell and electrode boundaries, the flux density maintains certain conditions. At the ends of the cell, there is no flux of ions, so the boundary condition is

$$N = 0 \text{ at } x = 0, L.\tag{3.35}$$

At internal boundaries between the electrodes and the separator we have the boundary condition

$$N|_{\delta-\epsilon}^{\delta+\epsilon} = 0,\tag{3.36}$$

where ϵ is infinitesimally small and δ is the interface location. This means that the flux density is continuous across the electrode–separator boundaries.

Unary Electrolyte

For insertion electrodes and a unary electrolyte, one ion of charge z and stoichiometry s produces n electrons that flow through the solid phase to the current collector. Conservation of charge in the reaction (Equation (3.2)) requires that $sz = -n$. For Ni–MH cells, $s = 1$, $z = -1$, and $n = 1$. For Li-ion cells, $s = -x$, $z = 1$, and $n = x$, where x depends on the composition of the lithiated carbon electrode. In Equation (3.34) the ion flux density

$$N = -D_e^{\text{eff}} \frac{\partial c}{\partial x},\tag{3.37}$$

where D_e^{eff} (cm^2/s) is the effective diffusion coefficient. The diffusion process represented by Equation (3.37) is caused by concentration gradients. The minimum energy state of a solution is attained when the species uniformly distributes throughout the volume. Gradients associated with a nonuniform species distribution force ions to travel in the direction of decreasing concentration.

The effective diffusion coefficient D_e^{eff} does not equal the diffusion coefficient of the bulk electrolyte D_e owing to the porosity of the electrode. The Bruggeman relation

$$D_e^{\text{eff}} = D_e \varepsilon \varepsilon^{0.5} = D_e \varepsilon^{1.5}\tag{3.38}$$

relates D_e^{eff} and D_e (assumed to be constant) with both having units of cm^2/s. The first ε in Equation (3.38) accounts for the porosity of the electrode and the second $\varepsilon^{0.5}$ term reduces the effective diffusion even further to account for tortuosity of the porous electrode. The circuitous route that the ions must travel through the electrode decreases the diffusion rate.

The ion flux density from the solid phase is

$$J = -\left(\frac{t_k}{z} + \frac{s}{n}\right)\frac{j}{F}\tag{3.39}$$

where $j(x, t)$ is given by the Butler–Volmer equation (3.19) and t_k is the transference number or fraction of the current carried by the ion. The two terms in Equation (3.39) represent the interfacial transfers of ions due to interface movement and microscopic diffusion, respectively. They result from averaging of the microscale concentration distributions in the porous structure to create a macroscale transport model. Roughly speaking, the current that fluxes from the solid to the electrolyte phase creates moles of ions that can be transported through the electrolyte.

The ion transport equation for Li-ion and Ni–MH cells can now be written using Equations (3.34)–(3.39). For Li-ion cells,

$$\varepsilon \frac{\partial c}{\partial t} = D_e^{\text{eff}} \frac{\partial^2 c}{\partial x^2} + a_s \left(\frac{1 - t_+}{F} \right) j, \tag{3.40}$$

where $c(x, t)$ is the concentration of lithium ions. For Ni–MH cells,

$$\varepsilon \frac{\partial c}{\partial t} = D_e^{\text{eff}} \frac{\partial^2 c}{\partial x^2} + a_s \left(\frac{t_- - 1}{F} \right) j, \tag{3.41}$$

where $c(x, t)$ is the concentration of hydroxide ions.

Binary Electrolyte

For a binary electrolyte, two ions of charge z_+ and z_- with stoichiometries s_+ and s_- produce n electrons that flow through the solid phase to the current collector. Conservation of charge in the reaction (Equation (3.2)) requires

$$s_+ z_+ + s_- z_- = -n. \tag{3.42}$$

For Pb–acid cells, we have different reactions in the positive and negative electrodes. In the negative (Pb) electrode, we have

$$\text{Pb} + \text{HSO}_4^- \rightleftharpoons \text{PbSO}_4 + \text{H}^+ + 2\text{e}^-, \tag{3.43}$$

so $s_- = 1, z_- = -1, s_+ = -1, z_+ = 1$, and $n = 2$. In the positive (PbO$_2$) electrode,

$$\text{PbO}_2 + \text{HSO}_4^- + 3\text{H}^+ + 2\text{e}^- \rightleftharpoons \text{PbSO}_4 + 2\text{H}_2\text{O}. \tag{3.44}$$

so $s_- = -1, z_- = -1, s_+ = -3, z_+ = 1$, and $n = 2$.

The ion flux density in Equation (3.34) is the same as for a unary electrolyte:

$$N = -D_e^{\text{eff}} \frac{\partial c}{\partial x} \tag{3.45}$$

where D_e^{eff} is the effective diffusion coefficient. For a binary electrolyte, the bulk diffusion coefficient D_e depends on the diffusion rates of both the negatively and positively charged species. One would expect that the overall diffusion rate would depend on an average of the two diffusion rates D_- and D_+. If the diffusion coefficients are different then the species are drawn apart by the application of an electric field. This polarizes the electrolyte and the species are drawn back together. This balance means that the faster ion pulls along the slower ion and the overall diffusion coefficient depends on both diffusion coefficients as follows:

$$D_e = \frac{(z_+ - z_-) D_+ D_-}{z_+ D_+ - z_- D_-} \tag{3.46}$$

where the Nernst–Einstein relation has been used. In the case of Pb–acid cells, Equation (3.46) reduces to

$$D_e = 2\frac{D_+D_-}{D_+ + D_-},\tag{3.47}$$

the familiar parallel impedance law with the factor of 2 giving $D_e = D_+ = D_-$ if the diffusion coefficients for the anion and cation are equal. To account for electrode porosity, the Bruggeman relation (3.38) is again used to determine D_e^{eff}.

The ion flux density from the solid phase is

$$J = -\left(\frac{t_+}{z_+} + \frac{t_-}{z_-} + \frac{s_+ + s_-}{n}\right)\frac{j}{2F}\tag{3.48}$$

where $j(x, t)$ is given by the Butler–Volmer equation (3.19) and t_+ and t_- are the transference numbers of the cation and anion, respectively. This equation generalizes Equation (3.39) for the binary electrolyte. In a binary electrolyte, the charge is carried by either the cations or the anions, so

$$t_+ = 1 - t_-.\tag{3.49}$$

The transference number can also be related to the diffusion coefficients, reducing the number of independent parameters in the model:

$$t_+ = \frac{z_+D_+}{z_+D_+ - z_-D_-}.\tag{3.50}$$

For Pb–acid cells, Equation (3.48) becomes

$$J = (1 - 2t_+)\frac{j}{2F}\tag{3.51}$$

in the negative electrode and

$$J = (3 - 2t_+)\frac{j}{2F}\tag{3.52}$$

in the positive electrode.

3.4.2   Conservation of Charge

The electrolyte, like the solid phase, is conductive and charge can flow, generally without a uniform distribution. In porous electrodes, the charge concentration or electric field in the electrolyte phase is driven by the current fluxing to and from the solid phase. The electrolyte allows charged particles to diffuse, adding a term to the solid phase model in Equation (3.38) that couples to the concentration. The charge carriers in the electrolyte are ions, as opposed to the electrons that carry charge in the solid phase.

Charge conservation in the electrolyte is expressed as

$$\kappa^{\text{eff}} \frac{\partial^2 \phi_e}{\partial x^2} + \kappa_d^{\text{eff}} \frac{\partial^2 c}{\partial x^2} = -a_s j, \tag{3.53}$$

where $\phi_e(x, t)$ is the potential of the electrolyte, $\kappa^{\text{eff}} = \kappa \varepsilon^{1.5}$ using the Bruggeman relation, κ (S/cm) is the electrolyte conductivity, $\kappa_d^{\text{eff}} = \kappa_d \varepsilon^{1.5}$, and κ_d (Acm^2/mol) is a constant. The first term in Equation (3.53) results from electrostatics in the (assumed uniform) conductive medium. The second term accounts for diffusion of the charged particles associated with concentration gradients. The distribution of charged particles throughout the domain affects the electric field ($\partial \phi_e / \partial x$) and the potential distribution. The final term is the flux of current from the solid phase. Note that this term is exactly the same as in Equation (3.38) except with the opposite sign. All of the current fluxing from the solid phase goes into the electrolyte phase and vice versa.

The electrolyte saturates the negative electrode, separator, and positive electrode of a battery cell. Thus, unlike the solid-phase potential (which is only defined in the porous electrodes but not the separator), the electrolyte potential is defined across the whole cell domain. The electrolyte, however, does not connect directly to the current collectors, so potential flux only comes through interaction with the active material.

Boundary conditions for the electrolyte potential are required at the two ends of the domain and at the interfaces between domains with different parameters (i.e., electrode–separator interfaces). The boundary conditions are related to the electric field in the electrolyte. At the ends of the domain, the electric field in the electrolyte is zero, so

$$\frac{\partial \phi_e}{\partial x} = 0 \text{ at } x = 0, L. \tag{3.54}$$

At the interface between an electrode and the separator the electric field is preserved, so

$$\left(\kappa^{\text{eff}} \frac{\partial \phi_e}{\partial x} + \kappa_d^{\text{eff}} \frac{\partial c}{\partial x} \right) \Big|_{\delta - \epsilon}^{\delta + \epsilon} = 0, \tag{3.55}$$

where the interface is located at $x = \delta$ and ϵ is infinitesimally small. The electrolyte has the same conductivity κ on both sides of the interface. The porosity, however, is not necessarily the same and, according to the Bruggeman relation, κ^{eff} may change from one side of the interface to the other. If the electrode and separator porosity are the same then Equation (3.55) simply requires continuity of the electric field.

Unary Electrolyte

The constants in Equation (3.53) can be related to other, previously introduced constants for unary and binary electrolytes. The conductivity of the unary electrolyte,

$$\kappa = \frac{z F^2 D_e}{RT} c, \tag{3.56}$$

is proportional to the diffusion coefficient and concentration of the electrolyte. Concentration varies with time and position within the cell. Thus, the first term in Equation (3.53) is nonlinear with a product of electrolyte potential and concentration. Linearization of Equation (3.53) simply requires that the average concentration be used in Equation (3.56), rendering κ constant. If the concentration varies widely (e.g., during deep discharge) then the assumption of constant electrolyte conductivity may not be accurate. The diffusion-related constant

$$\kappa_d = z F D_e \tag{3.57}$$

is also proportional to the diffusion coefficient of the electrolyte.

Binary Electrolyte

For a binary electrolyte, the conductivity depends on the mobility of both positive and negative ions as follows:

$$\kappa = \frac{v_+ z_+ F^2}{RT} \left(z_+ D_+ - z_- D_- \right) c, \tag{3.58}$$

where the quantity in parentheses is an averaged mobility. The conductivity of the binary electrolyte in Equation (3.58) is again proportional to concentration, rendering Equation (3.53) nonlinear. In most cases, however, it is reasonable to linearize Equation (3.58) by substituting $c(x, t) = c_{avg}$, the average concentration. The diffusion-related constant

$$\kappa_d = z_+ F \left(v_+ D_+ - v_- D_- \right) \tag{3.59}$$

is proportional to diffusion with different averaging weights (v_i) versus those in Equation (3.58) (z_i).

For Pb–acid cells, Equation (3.58) simplifies to

$$\kappa = \frac{F^2}{RT} \left(D_+ - D_- \right) c \tag{3.60}$$

and Equation (3.59) to

$$\kappa_d = F \left(D_+ - D_- \right). \tag{3.61}$$

3.4.3 Concentrated Solution Theory

The formulas relating diffusion coefficients and conductivity to the fundamental parameters of the reactions and the electrolyte work only for infinitely dilute solutions. For concentrated electrolytes these relationships may not apply . A detailed discussion of concentrated solution theory is beyond the scope of this text. Interested readers are referred to [11] and the references incorporated therein for further information on this topic.

The formulas provided for the parameters are a useful starting point for model tuning to match experimental data. The relative magnitudes are usually maintained, whereas the absolute values may vary significantly from the dilute theory predictions. In practice, even with the most advanced solution theories, it is very difficult to derive the model parameters from independent measurements. Research continues in this area to provide these critical measurements for battery model validation.

3.5   Cell Voltage

The current collectors connect to the solid electrodes at the ends of the domain ($x = 0$ and L). The solid-phase potential varies with position in the electrode. The output voltage of the cell is the difference between the potential at the positive current collector and the negative current collector, or

$$V(t) = \phi_s(L, t) - \phi_s(0, t) - \frac{R_f}{A} I, \qquad (3.62)$$

where R_f (Ω cm^2) is the contact resistance between the tab, current collectors, and the solid-phase electrode layer.

The voltage drop and rise within the cell is shown schematically in Figure 3.5. The overall cell voltage, excluding contact resistance, is the solid-phase potential in the positive electrode minus the solid-phase potential in the negative electrode, as shown in Equation (3.62). The voltage builds up through the cell from electronic and ionic resistances that cause voltage drops and open-circuit potential that can cause the voltage to rise or fall, depending on the sign and whether it occurs in the anode or cathode. Starting from the negative electrode, solid-phase resistance causes the voltage drop ΔV_s^-. The open-circuit potential is positive and, because it is in the negative electrode, the voltage drops by ΔV_U. The SEI kinetics create overpotential that causes a voltage drop of ΔV_η^-. The electrolyte ohmic resistance in the separator drops the voltage by ΔV_e. In the positive electrode, the charge transfer resistance and solid-phase ohmic resistance reduce the voltage by ΔV_η^+ and ΔV_s^+. The open-circuit potential in the positive

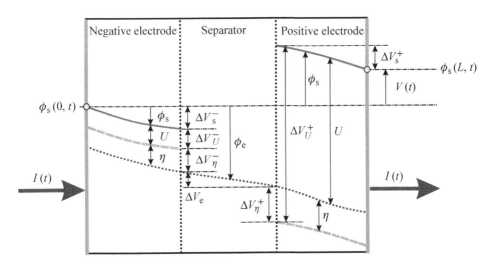

Figure 3.5   Sources of voltage drop and rise in the cell

electrode provides the driving force for current flow with a rise in voltage of ΔV_U^+. Overall the voltage can be written as

$$V(t) = \Delta V_U^+ - \Delta V_s^- - \Delta V_U^- - \Delta V_\eta^- - \Delta V_e - \Delta V_s^+ - \Delta V_\eta^+. \tag{3.63}$$

3.6 Cell Temperature

The performance and aging of batteries depend critically on temperature. Diffusion processes slow down at low temperatures. Side reactions can become dominant at extreme temperatures. Batteries are energy-dense devices and certain chemistries, under the right conditions, can exhibit thermal runaway where the temperature increases rapidly (self-heating rates above approximately 10 °C/min) and the pack can combust or explode. Battery system engineering requires careful evaluation of the effects of low and high temperatures on pack performance. Heating and/or cooling channels are often integrated into the pack to limit temperature extremes. Accurate modeling that includes temperature effects can predict performance, safety, and aging at extreme temperatures to facilitate the pack design process. The focus in this section is the thermal and electrochemical energy balance of the cell and the effects of temperature on cell parameters. Accurate simulation of a multicell pack, however, may require high-fidelity thermal models with the actual pack geometry, integrated heating/cooling, and boundary conditions. The models developed in this section can be used as source terms for a more complicated pack thermal model. Thermal-electrochemical coupled modeling is critical for simulations involving overcharge or high-rate charge/discharge that produce large overpotentials and substantial heating.

3.6.1 Arrhenius Equation

Many battery-cell model parameters depend on temperature T, including kinetic rate constants and transport properties. An Arrhenius dependence of these properties on temperature is often used [12]. For a temperature-dependent parameter Ψ, the Arrhenius relation is

$$\Psi = \Psi_{\text{ref}} \exp\left[\frac{E_{\text{act}}^\Psi}{R} \left(\frac{1}{T_{\text{ref}}} - \frac{1}{T} \right) \right], \tag{3.64}$$

where Ψ_{ref} is the property value defined at the reference temperature $T_{\text{ref}} = 25$ °C. The activation energy E_{act}^Ψ (J/mol) controls the temperature sensitivity of each individual property Ψ.

3.6.2 Conservation of Energy

As a first approximation, the temperature is assumed to be uniform throughout the cell at $T(t)$. Conservation of energy for this lumped-parameter cell determines the time evolution of temperature as follows:

$$C_p \frac{dT}{dt} = -hA_s(T - T_\infty) + q_i + q_j + q_c + q_r, \tag{3.65}$$

where h (W/(m^2 K)) is the heat transfer coefficient for forced convection, A_s is the cell surface area exposed to the convective cooling medium (typically air), T_∞ is the free stream temperature of the cooling medium, and C_p (J/K) is the heat capacity of the cell. Irreversible heat (watts) from the electrochemical reaction q_i, ohmic (Joule) heating q_j, reversible entropic heat q_r, and contact resistance q_c drive the cell temperature.

Volume-specific reaction heat generated in a finite control volume is equal to reaction current $j(x, t)$ times overpotential $\eta(x, t)$. The total reaction and Joule heat, $q_i + q_j$, is calculated by integrating the local volume-specific reaction heat across the one-dimensional cell domain and multiplying by plate area A or

$$q_i + q_j = A \int_0^L j \left(\phi_s - \phi_e - U \right) dx. \tag{3.66}$$

Note that there is no reaction, and thus no reaction heat generated, in the separator region. Reversible entropic heat

$$q_r = - \left(T \frac{\partial U}{\partial T} \right) I. \tag{3.67}$$

Additional ohmic heat arises from a contact resistance R_f between current collectors and electrodes. The total heat generated in the cell due to contact resistance is

$$q_c = I^2 \frac{R_f}{A}. \tag{3.68}$$

Since contact resistance R_f represents an empirical parameter in an otherwise fundamentally based model we list q_c separately from the previously mentioned ohmic heats q_j in Equation (3.66).

3.7   Side Reactions and Aging

Side reactions occur in all cells under the right conditions. Some side reactions are benign, being completely reversible with no long-term effects. Other side reactions are not completely reversible and cause permanent degradation to the cell. These are the side reactions that govern aging or the slow decay of a battery cell's capacity. In this section we model the primary capacity-degrading side reactions in Li-ion cells. The actual degradation mechanism for a particular cell could be different, however. Experimental, long-term cycle testing and forensic analysis of degraded cells is the best method to determine the aging mechanism for a specific battery.

In Li-ion cells, side reactions that may cause aging occur in both the anode and cathode [13]. For the carbon-based anode, a layer forms at the SEI layer. This layer is beneficial and essential for proper cell operation, but an electrolyte reduction side reaction can occur that causes aging. In this side reaction the electrolyte (e.g., ethylene carbonate) reacts to increase the thickness of the SEI layer due to product precipitation. A surface film on the active particles also occurs in the cathode. The film thickness does not change significantly during aging, but its porosity, conductivity, and diffusion coefficient do change over time due to precipitate of side-reaction product that blocks the pores of the existing surface film.

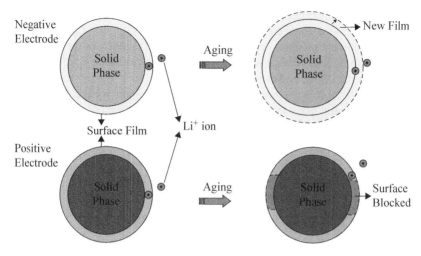

Figure 3.6 Aging mechanisms for the Li-ion cell: negative electrode (top) and positive electrode (bottom).

Figure 3.6 shows the two aging mechanisms for the Li-ion cell. In the negative electrode, the growing SEI layer increases the cell impedance and permanently removes lithium from the cell, thus reducing capacity. In the positive electrode, access to the active particles is restricted by precipitate, increasing impedance and reducing the active material and cell capacity.

The surface film model used in both the anode and cathode is shown in Figure 3.7. The solid particle radius is much larger than the film thickness, so planar diffusion equations are used

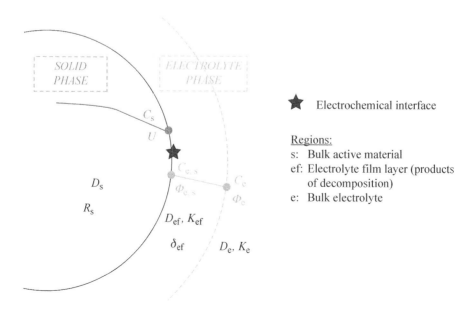

Figure 3.7   SEI layer model for Li-ion cell

for the lithium concentration $c_{ef}(r, t)$ and potential $\phi_{ef}(r, t)$, where the subscript "ef" stands for electrolyte film. The film concentration dynamics take the form

$$\frac{\partial c_{ef}}{\partial t} = D_{ef} \frac{\partial^2 c_{ef}}{\partial r^2} \tag{3.69}$$

and the potential distribution is governed by

$$\kappa_{ef} \frac{\partial^2 \phi_{ef}}{\partial r^2} + \kappa_{ef}^D \frac{\partial^2 c_{ef}}{\partial r^2} = 0 \tag{3.70}$$

for $r \in (R_s, R_s + \delta_{ef})$. The parameters D_{ef}, κ_{ef}, and κ_{ef}^D are assumed constant.

The side reaction current densities j_a^s (anode) and j_c^s (cathode) are assumed to have Butler–Volmer kinetics:

$$j_a^s = -a_{s,a}^s i_{0,a}^s \exp\left[-\frac{\alpha_a^s}{RT} \left(\phi_s - \phi_{e,s} - U_a^s \right) \right] \tag{3.71}$$

$$j_c^s = -a_{s,c}^s i_{0,c}^s \exp\left[-\frac{\alpha_c^s}{RT} \left(\phi_s - \phi_{e,s} - U_c^s \right) \right] \tag{3.72}$$

based on the model parameters $a_{s,a}^s$, $i_{0,a}^s$, α_a^s, U_a^s, $a_{s,c}^s$, $i_{0,c}^s$, α_c^s, and U_c^s. The current densities enter Equation (3.69) through a flux boundary condition. The solid-phase conservation equations (charge and species) in the baseline cell model use modified source terms $j + j_a^s$ for the anode and $j + j_c^s$ for the cathode.

The growth of the anode SEI layer is modeled by

$$\frac{\partial \delta_{ef,a}}{\partial t} = \frac{j_a^s}{a_{s,a} nF} \frac{M_a^s}{\rho_a^s}, \tag{3.73}$$

where M^s and ρ^s are the molecular weight and density of the product species, respectively. The loss rate of active lithium due to SEI layer formation is

$$\dot{c}_{s,lost}^s = \frac{1}{nF\varepsilon_s L} \int_0^L j_a^s \, dx. \tag{3.74}$$

In the cathode, the film porosity $\varepsilon_{ef,c}$ is gradually reduced during cycle life using a rate law:

$$\frac{\partial \varepsilon_{ef,c}}{\partial t} = -\frac{j_c^s}{a_{s,c} nF} \frac{M_c^s}{\rho_c^s}, \tag{3.75}$$

where M^s and ρ^s are the molecular weight and density of the product species, respectively. The cathode film conductivity and diffusion coefficient can then be adjusted during aging from reference values using Bruggeman-type relations. The cathode capacity is reduced during cycling because the active surface area per unit volume is decreasing due to side-reaction product precipitation as follows:

$$a_c = a_c^0 \left[1 - \left(\frac{\varepsilon_{ef,c}^0 - \varepsilon_{ef,c}}{\varepsilon_{ef,c}^0} \right)^\zeta \right], \tag{3.76}$$

where a_c^0 and $\varepsilon_{ef,c}^0$ are initial values and ζ is an experimentally determined parameter.

The Li-ion degradation model treats the anode SEI film as having constant film parameters except for a growing thickness. The cathode surface film parameters, on the other hand, are life dependent except for a constant thickness. For further information on the modeling of SEI layer formation and growth in Li-ion cells, readers are referred to [14, 15].

Problems

3.1   Calculate the theoretical specific energy density of an Ni–MH battery by assuming that all the reactants are fully joined in the electrochemical reaction

$$NiOOH + MH \xrightleftharpoons[\text{charge}]{\text{discharge}} Ni(OH)_2 + M. \tag{A.3.1}$$

The molecular weights for NiOOH, H_2O, Ni(OH)$_2$, MH, and M are 91.7 g/mol, 18 g/mol, 92.7 g/mol, 57.1 g/mol, and 56.1 g/mol, respectively.

3.2   Pb–acid cells have the half-reactions

$$PbO_2 + HSO_4^- + 3H^+ + 2e^- \xrightleftharpoons[\text{charge}]{\text{discharge}} PbSO_4 + 2H_2O, \tag{A.3.2}$$

$$Pb + HSO_4^- \xrightleftharpoons[\text{charge}]{\text{discharge}} PbSO_4 + H^+ + 2e^-. \tag{A.3.3}$$

Calculate the theoretical specific energy (J/kg) of a Pb–acid battery cell, using the molar masses $Pb = 207.2 \text{g/mol}$, $PbO_2 = 239.2 \text{g/mol}$, and $H_2SO_4 = 98.1 \text{g/mol}$.

3.3   Find the theoretical specific energies of Li-ion batteries with the following two chemistries:

(a)   LiCoO$_2$ cells

$$Li_{0.66}C_6 \rightleftharpoons 0.66Li^+ + 0.66e^- + 6C \quad \text{in the negative electrode,}$$

$$0.66Li^+ + 0.66e^- + Li_{0.34}CoO_2 \rightleftharpoons LiCoO_2 \quad \text{in the positive electrode.}$$

(b)   LiMnO$_2$ cells

$$Li_{0.66}C_6 \rightleftharpoons 0.66Li^+ + 0.66e^- + 6C \quad \text{in the negative electrode,}$$

$$0.66Li^+ + 0.66e^- + Li_{0.34}MnO_2 \rightleftharpoons LiMnO_2 \quad \text{in the positive electrode.}$$

3.4 Determine the units of each of the following equations:
   (a) Equation (3.26)
   (b) Equation (3.29)
   (c) Equation (3.40)
   (d) Equation (3.53).
   Example: Equation (3.16)

$$\dot{Q} = I\left(V - U + T\frac{\partial U}{\partial T}\right)$$

$$\frac{J}{s} = \frac{C}{s}\left(\frac{J}{C} - \frac{J}{C} + \frac{J}{C}\right)$$

$$\frac{J}{s} = \frac{J}{s}.$$

Answer: Units = J/s.

3.5 Integrate the conservation of charge equation for an electrode

$$\sigma^{\text{eff}}\frac{\partial^2\phi_s}{\partial x^2} = a_s j, \tag{A.3.4}$$

with boundary conditions

$$\frac{\partial\phi_s}{\partial x}(0, t) = 0 \tag{A.3.5}$$

and

$$\sigma^{\text{eff}}\frac{\partial\phi_s}{\partial x}(L, t) = \frac{I}{A} \tag{A.3.6}$$

to obtain a relationship between the average current density

$$j_{\text{avg}} = \frac{1}{L}\int_0^L j(x, t)\,dx \tag{A.3.7}$$

and the input current $I(t)$.

3.6 Diffusion in a particle is governed by

$$\frac{\partial c_s}{\partial t} = D_s\left(\frac{\partial^2 c_s}{\partial r^2} + \frac{m}{r}\frac{\partial c_s}{\partial r}\right) \tag{A.3.8}$$

where $m = 2$ for spherical Li-ion particles and $m = 1$ for cylindrical Ni–MH particles. The boundary conditions are

$$\left.\frac{\partial c_s}{\partial r}\right|_{r=0} = 0, \tag{A.3.9}$$

and

$$\left.\frac{\partial c_s}{\partial r}\right|_{r=R_s} = \frac{-j}{D_s F}. \tag{A.3.10}$$

Determine a differential equation for the volume-averaged concentration

$$c_{s,avg} = \frac{1}{V} \int_0^{R_s} c_s \, dV. \tag{A.3.11}$$

for (a) an Li-ion particle and (b) an Ni–MH particle.

3.7 Consider ion transport in an electrolyte governed by

$$\varepsilon \frac{\partial c}{\partial t} = D \frac{\partial^2 c}{\partial x^2} + \frac{ab}{F} j \tag{A.3.12}$$

in two domains, $x \in (0, \delta)$ and $x \in (\delta, L)$, corresponding to a negative and positive electrode, respectively. In each domain, the parameters a, b, ε, and D have constant values, but the values in the negative electrode (a_m, b_m, ε_m, D_m) are different than those in the positive electrode (a_p, b_p, ε_p, D_p). The boundary conditions are

$$\frac{\partial c}{\partial x} = 0 \quad \text{at } x = 0, L, \tag{A.3.13}$$

$$c(\delta^-, t) = c(\delta^+, t), \tag{A.3.14}$$

$$D_m \frac{\partial c}{\partial x}(\delta^-, t) = D_p \frac{\partial c}{\partial x}(\delta^+, t). \tag{A.3.15}$$

(a) Determine the average concentration dynamics for the cell where

$$c_{avg} = \frac{1}{L} \int_0^L c(x, t) \, dx. \tag{A.3.16}$$

given the average current density

$$j_{m,avg} = \frac{1}{\delta} \int_0^\delta j(x, t) \, dx = \frac{I(t)}{a_m A \delta}, \tag{A.3.17}$$

$$j_{p,avg} = \frac{1}{L - \delta} \int_\delta^L j(x, t) \, dx = -\frac{I(t)}{a_p A(L - \delta)}. \tag{A.3.18}$$

A.3.16 What conclusions can you draw about the average concentration in an Li-ion cell where $b_m = b_p$ and a Pb–acid cell where b_m and b_p have opposite sign? Assume all other parameters are positive.

3.8 Figure A.3.1 shows a simplified equivalent circuit of a single battery cell electrode.
   (a) Derive the terminal voltage V as a function of given current I and cell potential U.

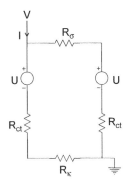

Figure A.3.1   Equivalent circuit of a single battery cell electrode

(b)   If a battery is discharged too rapidly the potential can drop quickly to zero in part
of the electrode. Determine the terminal voltage V if one of the potentials U is
zero.

3.9   For an Li-ion cell, find an expression for the average lithium concentration in the
electrolyte for the positive electrode given by

$$c_{e,p,avg} = \frac{1}{\delta_p} \int_0^{\delta_p} c_e \, dx \tag{A.3.19}$$

by integrating the electrolyte diffusion equation

$$\varepsilon \frac{\partial c_e}{\partial t} = D_e^{eff} \frac{\partial^2 c_e}{\partial x^2} + a_s \frac{1 - t_+}{F} j \tag{A.3.20}$$

with boundary conditions

$$\left. \frac{\partial c_e}{\partial x} \right|_{x=0} = 0, \tag{A.3.21}$$

$$\left. \frac{\partial c_e}{\partial x} \right|_{x=\delta_p} = -\frac{c_e(\delta_p)}{\delta_{sep}/2}. \tag{A.3.22}$$

The boundary condition (A.3.22) at the interface between the positive electrode and
separator results from assuming a linear concentration distribution within the separator
and zero concentration at the midpoint of the separator. This is a reasonable approximation for many Li-ion cells because the ion production rates in the two electrodes are
opposite in sign and roughly equal in magnitude, resulting in an electrolyte concentration distribution that is approximately antisymmetric about the middle of the separator,
as shown in Figure A.3.2.

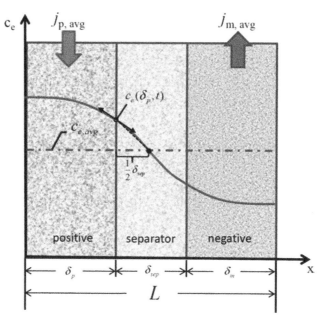

Figure A.3.2   Symmetric electrolyte diffusion model

3.10   Electrolyte charge conservation is

$$\kappa^{\text{eff}}\frac{\partial^2 \phi_e}{\partial x^2} + k_d^{\text{eff}}\frac{\partial^2 c}{\partial x^2} = -a_s j, \tag{A.3.23}$$

where the concentration $c(x, t)$ and potential $\phi_e(x, t)$ have zero flux boundary conditions at the ends of the cell $x = 0, L$. Calculate the overall cell conservation of charge by calculating

$$j_{\text{avg}} = \frac{1}{L}\int_0^L j(x, t)\,\mathrm{d}x. \tag{A.3.24}$$

3.11   Linearize the Arrhenius equation (3.64) at $T = T_{\text{ref}}$.

4

Discretization Methods

The governing PDEs derived in Chapter 3 constitute the building blocks of battery models. To be useful to systems engineers, however, these equations must be discretized in space to reduce them to multiple, ordinary differential equations (ODEs) in time. In this chapter we introduce several methods that can be used to discretize the governing equations to standard state-variable and transfer-function forms. The state-variable form is

$$\dot{\mathbf{x}}(t) = \mathbf{A}\mathbf{x}(t) + \mathbf{B}\mathbf{u}(t), \tag{4.1a}$$

$$\mathbf{y}(t) = \mathbf{C}\mathbf{x}(t) + \mathbf{D}\mathbf{u}(t), \tag{4.1b}$$

where $\dot{\mathbf{x}} = d\mathbf{x}/dt$, $\mathbf{x}(t) \in R^N$ is the state vector, $\mathbf{u}(t) \in R^M$ is the input vector, $\mathbf{y}(t) \in R^P$ is the output vector, $\mathbf{A} \in R^{N \times N}$ is the state matrix, $\mathbf{B} \in R^{N \times M}$ is the input matrix, $\mathbf{C} \in R^{P \times N}$ is the output matrix, and $\mathbf{D} \in R^{P \times M}$. Here, N is the number of states, M is the number of inputs (typically only one input for batteries – current), and P is the number of outputs (including the voltage). The standard transfer-function form for a multi-input multi-output system is

$$\mathbf{Y}(s) = \mathbf{G}(s)\mathbf{U}(s) \tag{4.2}$$

where $\mathbf{Y}(s)$ and $\mathbf{U}(s)$ are the Laplace transforms of $\mathbf{y}(t)$ and $\mathbf{u}(t)$, respectively, and $\mathbf{G}(s) \in R^{P \times M}$ is the transfer function matrix.

In this chapter we will study analytical and numerical methods for discretizing the governing equations of battery models. Analytical methods are typically restricted to linear problems with constant coefficients or just a few coupled domains with constant coefficients. In some special cases, solutions can be found for systems with spatially varying coefficients or simple nonlinearities. For nonlinear or non-constant coefficient PDEs, analytical solutions may be difficult or impossible to obtain. Numerical methods must therefore be used to discretize the governing equations. Six methods are studied: integral method approximation (IMA), Padé approximation, Ritz method, finite-element method (FEM), finite-difference method (FDM), and system identification method. The IMA uses polynomial approximations of the field variables over the domain. Integration of the field equation and the boundary conditions is used to determine the approximate solution. The Padé approximation method converts

Battery Systems Engineering, First Edition. Christopher D. Rahn and Chao-Yang Wang.
© 2013 John Wiley & Sons, Ltd. Published 2013 by John Wiley & Sons, Ltd.

transcendental transfer functions (the exact solutions of diffusion equations) into polynomial transfer functions that are used in systems engineering. The Ritz method and FEM use a variational form of the governing equations that incorporates all of the natural (i.e., flux) boundary conditions. In the Ritz method, expansion functions (polynomial and Fourier) are defined across the entire domain. For the FEM, the domain is divided into elements with linear interpolation functions. In both cases, the solutions exhibit smooth convergence properties and maintain the symmetry of the underlying operators. The FDM also divides the domain into elements and approximates the spatial derivatives using simple difference formulas. The system identification method is a numerical technique for generating a low-order polynomial transfer function from a high-order or transcendental one.

The examples in this chapter come from simplified battery models that can be used as a basis for further model development, including the complete battery models developed in Chapter 6. Models of electrolyte diffusion, solid-state diffusion, and coupled electrolyte–solid-phase diffusion are studied. Most of the models are solved using more than one method to compare and contrast the different approaches. In Chapter 5, many of these models will be simulated to determine accuracy and numerical efficiency. In particular, the coupled domains electrolyte diffusion model will be developed in this chapter and simulated in Chapter 5 for all of the solution methods to compare their response predictions and calculation speeds. For real-time implementation, an accurate low-order model is required and the various techniques will be compared to determine what model order is necessary to obtain a specified response accuracy under prescribed loading conditions.

4.1 Analytical Method

Most of the PDEs that are encountered in battery systems are approximately linear with constant coefficients so we can often find an exact or analytical solution. In this section, we solve electrolyte and solid-phase diffusion problems that are representative of battery systems. More information on analytical methods can be found in [16–19].

Two approaches are used to analytically/exactly solve the PDEs. First, we use the separation of variables to generate an eigenvalue problem that is then solved. The spatially distributed response is calculated from an eigenfunction series expansion. The electrolyte has concentration distributions across the cell width that are of interest, so an eigenfunction approach is taken for one-dimensional planar diffusion, first for a single domain and then two coupled domains. Second, we use the Laplace transform to eliminate time derivatives and solve the resulting ODEs for a transcendental transfer function. The transfer function does not provide spatial distributions, so it is typically used when only the output is needed. We will introduce the transfer function approach with the planar electrolyte model and use it exclusively for cylindrical and spherical particle models.

4.1.1 Electrolyte Diffusion

Single Domain

Figure 4.1 shows a single-domain model of diffusion that arises from conservation of species in the electrolyte of Pb–acid, Li-ion, and Ni–MH batteries. The governing equation is

$$\frac{\partial c}{\partial t} = a_1 \frac{\partial^2 c}{\partial x^2} + a_2 j \quad \text{for } x \in (0, L), \tag{4.3}$$

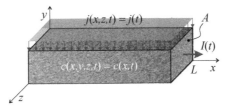

Figure 4.1 Electrolyte phase diffusion problem for a porous electrode

where $c(x, t)$ is concentration. The constants $a_1 = D/\varepsilon$ and $a_2 = (1 - t^0)/(\varepsilon F)$, where D is the diffusion coefficient, ε is the electrolyte phase volume fraction, t^0 is the transference number, and F is the Faraday constant. The transfer current density $j(t)$ is assumed to be uniformly distributed throughout the domain $0 < x < L$. The boundary conditions are

$$\left.\frac{\partial c}{\partial x}\right|_{x=0} = \left.\frac{\partial c}{\partial x}\right|_{x=L} = 0. \tag{4.4}$$

An electrode at $x = L$ provides a concentration flux and changes the boundary condition to

$$\left.\frac{\partial c}{\partial x}\right|_{x=L} = a_3 I(t), \tag{4.5}$$

where $I(t)$ is a current input and $a_3 = 1/(AF)$ with A the electrode plate area. Initial conditions on the concentration complete the model:

$$c(x, 0) = c_{\text{ic}}(x). \tag{4.6}$$

Separation of variables by substituting $c(x, t) = C(x) e^{\lambda t}$ into Equation (4.3) with $j = 0$ produces

$$C\lambda e^{\lambda t} = a_1 C'' e^{\lambda t} \quad \forall t > 0, \tag{4.7}$$

where we use the shorthand notation $C' = dC/dx$. Equation (4.7) simplifies to

$$\lambda C - a_1 C'' = 0. \tag{4.8}$$

Equation (4.8) is an eigenvalue problem. We seek eigenfunctions $C(x) \neq 0$ and eigenvalues λ that satisfy Equation (4.8). Equation (4.8) is linear, so exponential solutions of the form $C(x) = e^{\beta x}$ work if $\lambda = a_1 \beta^2$. Thus, a solution of the form

$$C(x) = C_1 e^{\beta x} + C_2 e^{-\beta x} \tag{4.9}$$

satisfies the eigenvalue problem (4.8) with $\beta = \sqrt{\lambda/a_1}$.

Transfer Current Input

For the single-domain problem defined by the field equation (4.3), we find eigenvalues λ and eigenfunction $C(x)$ that satisfy the boundary conditions (4.4) as follows:

$$\frac{\partial c}{\partial x}\Big|_{x=0} = C_1\beta - C_2\beta = 0$$

$$\frac{\partial c}{\partial x}\Big|_{x=L} = C_1\beta\, e^{\beta L} - C_2\beta\, e^{-\beta L} = 0 \tag{4.10}$$

Simultaneous solution of Equation (4.10) yields $C_1 = C_2$ and

$$e^{\beta L} - e^{-\beta L} = 0. \tag{4.11}$$

Equation (4.11) has solutions if $\beta = \sqrt{\lambda/a_1}$ is purely imaginary or if $\lambda < 0$. In this case,

$$e^{i\sqrt{|\lambda|/a_1}\,L} - e^{-i\sqrt{|\lambda|/a_1}\,L} = \cos(\sqrt{|\lambda|/a_1}\,L) + i\sin(\sqrt{|\lambda|/a_1}\,L)$$
$$- \cos(\sqrt{|\lambda|/a_1}\,L) + i\sin(\sqrt{|\lambda|/a_1}\,L) \tag{4.12}$$
$$= 2i\sin(\sqrt{|\lambda|/a_1}\,L) = 0.$$

The eigenvalue solutions are $\sqrt{|\lambda_m|/a_1}\,L = m\pi$ or $\lambda_m = -a_1(m\pi/L)^2$ for $m = 0, 1, \ldots$. Thus, the eigenfunctions are

$$C_0(x) = \frac{1}{\sqrt{L}}, \quad C_m(x) = \sqrt{\frac{2}{L}}\cos(m\pi x/L), \tag{4.13}$$

where they have been normalized such that

$$\int_0^L C_m^2\, dx = 1. \tag{4.14}$$

The eigenfunctions are orthogonal as well because

$$\int_0^L C_m C_n\, dx = 0 \quad \text{if } m \neq n. \tag{4.15}$$

Another orthogonality relation is

$$\int_0^L a_1 C_m'' C_n\, dx = \lambda_m \int_0^L C_m C_n\, dx = \lambda_m \delta_{mn}, \tag{4.16}$$

where $\delta_{mm} = 1$ and $\delta_{mn} = 0$ if $m \neq n$. In Equation (4.16) we have used the eigenvalue problem (4.8).

The exact solution of Equation (4.3) with boundary conditions (4.4) is an infinite eigen-function series

$$c(x, t) = \sum_{m=0}^{\infty} C_m(x)c_m(t).$$ (4.17)

Substitution of Equation (4.17) into Equation (4.3), multiplication by $C_n(x)$, and integration yields

$$\int_0^L \sum_{m=0}^{\infty} C_m C_n \dot{c}_m \, dx = \int_0^L \left(\sum_{m=0}^{\infty} a_1 C_m'' c + a_2 j \right) C_n \, dx.$$ (4.18)

Using the orthogonality Equations (4.15) and (4.16), Equation (4.18) simplifies to

$$\dot{c}_n(t) = \lambda_n c_n(t) + b_n j(t) \quad \text{for } n = 1, \ldots,$$ (4.19)

where

$$b_n = a_2 \int_0^L C_n \, dx = 0 \quad \text{for } n > 0$$ (4.20)

and $b_0 = a_2 \sqrt{L}$.

Equation (4.19) can be written in state-space form (4.1) with

$$\mathbf{x}(t) = [c_0(t), c_1(t), \ldots, c_N(t)]^{\mathrm{T}},$$ (4.21)

$$\mathbf{A}(m, n) = \lambda_m \delta_{mn}, \quad \mathbf{B}(m) = b_m,$$ (4.22)

$$\mathbf{C}(n) = C_n(L), \quad \mathbf{D} = 0.$$ (4.23)

The input $\mathbf{u}(t) = j(t)$ and the output

$$y(t) = c(L, t) = \sum_{n=0}^{N} C_n(L) c_n(t).$$ (4.24)

We must select the approximation order N to realize the discretized state-space model. The transfer function is obtained by taking the Laplace transform of Equation (4.19) for $n = 0$ ($c_n(t)$ for $n > 0$ are not excited by the input) to produce

$$\frac{Y(s)}{U(s)} = \frac{a_2}{s}$$ (4.25)

so the system simply acts as an integrator. Although the transfer function is simple, it cannot be used to simulate the response to initial conditions. The initial conditions on $c(x, t)$ in Equation (4.6) correspond to initial conditions on the state, $\mathbf{x}(0)$, as follows:

$$\int_0^L C_n(x)c(x, 0)\,dx = \int_0^L C_n(x) \sum_{m=0}^{\infty} C_m(x)c_m(t)\,dx = \int_0^L C_n(x)C_{ic}(x)\,dx \quad (4.26)$$

where we have premultiplied Equation (4.17) by $C_n(x)$ and integrated over the domain. Using the orthogonality relations (4.15), we obtain the initial conditions on the state

$$c_n(0) = \int_0^L C_n(x)C_{ic}(x)\,dx. \quad (4.27)$$

Boundary Current Flux Input
Changing the boundary conditions to the boundary flux input (4.5) does not change the eigenvalue problem. Eigenvalue problems concern the unforced response, so we set all the inputs to zero. Thus, $I(t) = 0$, giving the same zero-flux boundary condition as in the previous case. Following the solution process of Equations (4.17)–(4.21) would give the same $\mathbf{A}$, $\mathbf{B}$, $\mathbf{C}$, and $\mathbf{D}$ matrices, so the input $I(t)$ does not appear in the final equations. To correct this problem, we generate a weak form of Equation (4.3) by premultiplying by $C_n(x)$ and integrating over the domain to produce

$$\int_0^L C_n(x)\dot{c}(x, t)\,dx = \int_0^L C_n(x)a_1 c''(x, t)\,dx, \quad (4.28)$$

where $\dot{c} = \partial c/\partial t$. Integration by parts brings the boundary condition (4.5) into Equation (4.28):

$$\int_0^L C_n \dot{c}\,dx = a_1 C_n c' \big|_0^L - a_1 \int_0^L C_n' c'\,dx + a_2 \int_0^L C_n\,dx\, j$$
$$= a_1 a_3 C_n(L)I(t) - a_1 \int_0^L C_n' c'\,dx. \quad (4.29)$$

Following steps (4.17)–(4.21) produces the same $\mathbf{A}$, $\mathbf{C}$, and $\mathbf{D}$ matrices as in Equation (4.21), but $\mathbf{u}(t) = I(t)$ and

$$\mathbf{B} = \begin{bmatrix} a_1 a_3 C_0(L) \\ \cdots \\ a_1 a_3 C_N(L) \end{bmatrix}. \quad (4.30)$$

The transfer function can be obtained by taking the Laplace transform of (4.19) including the input $I(t)$ and summing according to Equation (4.24) to produce

$$G(s) = \frac{C(L, s)}{I(s)} = \sum_{n=0}^{N} \frac{a_1 a_3 C_n^2(L)}{s + \lambda_n}. \quad (4.31)$$

We can obtain a transcendental version of $G(s)$ by applying the Laplace transform directly to Equation (4.3) with $j = 0$ to produce the ODE

$$sC(x, s) = a_1 C''(x, s). \tag{4.32}$$

As in the eigenvalue problem, Equation (4.32) has solutions of the form

$$C(x, s) = C_1(s)\, e^{\beta x} + C_2(s)\, e^{-\beta x}. \tag{4.33}$$

Substitution of Equation (4.33) into Equation (4.32) yields $\beta = \sqrt{s/a_1}$. We substitute Equation (4.33) into the Laplace transformed boundary conditions

$$C'(0, s) = C_1(s) - C_2(s) = 0,$$
$$C'(L, s) = \beta(C_1(s)\, e^{\beta L} - C_2(s)\, e^{-\beta L}) = a_3 I(s). \tag{4.34}$$

Simultaneous solution of (4.34) for $C_1(s)$ and $C_2(s)$ and substitution into Equation (4.33) evaluated at $x = L$ yields the transcendental transfer function

$$\frac{C(L, s)}{I(s)} = C_1(s)\, e^{\beta L} + C_2(s)\, e^{-\beta L} \tag{4.35}$$

$$= \frac{a_3 \cosh(\beta L)}{\beta \sinh(\beta L)}, \tag{4.36}$$

where $\beta = \sqrt{s/a_1}$.

We can convert the transcendental transfer function into the series expansion in Equation (4.31) by using the partial fraction expansion

$$G(s) = \sum_{n=0}^{\infty} \frac{\text{Res}_n}{s - \lambda_n} \tag{4.37}$$

if the poles are real and not repeated. The residues are calculated from

$$\text{Res}_n = \lim_{s \to \lambda_n} (s - \lambda_n) G(s) \tag{4.38}$$

The poles of $G(s)$ in Equation (4.36) are at $\beta = 0$ ($s = 0$) and

$$\sinh(\beta L) = \sinh\left(\sqrt{\frac{sL^2}{a_1}}\right) = 0 \tag{4.39}$$

Equation (4.39) has only negative real poles at $s = \lambda_n = -a_1(n\pi/L)^2$. Using Equation (4.38), the residues

$$\text{Res}_n(s) = a_1 a_3 C_n^2(L), \tag{4.40}$$

agreeing with the previously derived transfer function (4.31).

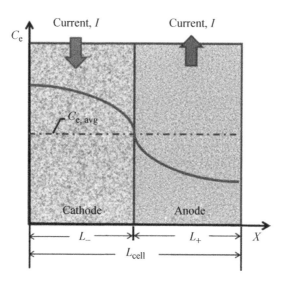

Figure 4.2 Electrolyte phase diffusion problem for Li-ion cell with uniform reaction current distribution

Coupled Domains

Figure 4.2 shows an electrolyte phase diffusion problem for a battery cell with uniform reaction current distribution and two coupled domains. The two domains correspond to a porous negative electrode $(0 < x < L/2)$ and a porous positive electrode $(L/2 < x < L)$ with an extremely thin separator between them to prevent the flow of electrons. We assume for simplicity that the two electrodes are the same length $(L/2)$ and that the diffusion coefficients and electrode phase volume fractions are different for (but constant within) the two electrodes. The current density $j(t) = 2I(t)/(AL)$ for the negative electrode and $j(t) = -2I(t)/(AL)$ for the positive electrode, where A is the electrode plate area and $I(t)$ is the total current flowing through the cell. Thus, the coupled-domains model consists of the two field equations

$$\varepsilon_m \frac{\partial c}{\partial t} = D_m \frac{\partial^2 c}{\partial x^2} + bI \quad \text{for } x \in (0, L/2), \tag{4.41}$$

$$\varepsilon_p \frac{\partial c}{\partial t} = D_p \frac{\partial^2 c}{\partial x^2} - bI \quad \text{for } x \in (L/2, L), \tag{4.42}$$

where ε_m, ε_p and D_m, D_p are the electrode phase volume fractions and diffusion coefficients for the negative and positive electrodes, respectively. The diffusion coefficients depend on a reference diffusion coefficient D^{ref} and the phase volume fractions as follows: $D_m = D^{ref}\varepsilon_m^{1.5}$ and $D_p = D^{ref}\varepsilon_p^{1.5}$. The input constant

$$b = \frac{2(1 - t_0)}{FAL}, \tag{4.43}$$

where t_0 is the transference number and F is the Faraday constant.

The boundary conditions at $x = 0$ and $x = L$ are

$$\frac{\partial c}{\partial x}\bigg|_{x=0} = \frac{\partial c}{\partial x}\bigg|_{x=L} = 0 \tag{4.44}$$

and at the interface between the two domains $x = L/2$ they are

$$D_m \frac{\partial c}{\partial x}\bigg|_{x=(L/2)_-} = D_p \frac{\partial c}{\partial x}\bigg|_{x=(L/2)_+}$$

$$c\left(\frac{L}{2}_-, t\right) = c\left(\frac{L}{2}_+, t\right) \tag{4.45}$$

The boundary conditions (4.45) couple the two domains by ensuring continuity of concentration and flux through the boundary at $x = L/2$.

Eigenfunction Expansion

The eigenvalue problem is derived by substituting $c(x, t) = C(x) e^{\lambda t}$ into Equations (4.41) and (4.42) with $I(t) = 0$ to produce

$$\varepsilon_m \lambda C_m - D_m C_m'' = 0 \quad \text{for } x \in (0, L/2), \tag{4.46}$$

$$\varepsilon_p \lambda C_p - D_p C_p'' = 0 \quad \text{for } x \in (L/2, L). \tag{4.47}$$

The solutions of Equations (4.46) and (4.47) are

$$C_m(x) = C_{1m} e^{\beta_m x} + C_{2m} e^{-\beta_m x}, \tag{4.48}$$

$$C_p(x) = C_{1p} e^{\beta_p x} + C_{2p} e^{-\beta_p x}. \tag{4.49}$$

Substitution of Equations (4.48) and (4.49) into Equations (4.46) and (4.47) produces

$$\lambda = \frac{D_m \beta_m^2}{\varepsilon_m} = \frac{D_p \beta_p^2}{\varepsilon_p} \tag{4.50}$$

or

$$\beta_m = \alpha \beta_p, \quad \varepsilon_m = \zeta \varepsilon_p, \text{ and } D_m = \frac{\zeta D_p}{\alpha^2} \tag{4.51}$$

where

$$\alpha = \sqrt{\frac{D_p \varepsilon_m}{D_m \varepsilon_p}} \text{ and } \zeta = \frac{\varepsilon_m}{\varepsilon_p}. \tag{4.52}$$

Substitution of Equation (4.50) into the solutions (4.48) and (4.49) and then into the boundary conditions (4.44) and (4.45) produces the matrix equation

$$\mathbf{Mc} = \mathbf{0},\tag{4.53}$$

where $\mathbf{c} = [C_{1m}, C_{2m}, C_{1p}, C_{2p}]^T$ and

$$\mathbf{M} = \begin{bmatrix} \alpha\beta_p & -\alpha\beta_p & 0 & 0 \\ \alpha\zeta\beta_p \, e^{\frac{1}{2}\alpha\beta_p L} & -\alpha\zeta\beta_p \, e^{-\frac{1}{2}\alpha\beta_p L} & -\alpha^2\beta_p \, e^{\frac{1}{2}\beta_p L} & \alpha^2\beta_p \, e^{-\frac{1}{2}\beta_p L} \\ e^{\frac{1}{2}\alpha\beta_p L} & e^{-\frac{1}{2}\alpha\beta_p L} & -e^{\frac{1}{2}\beta_p L} & -e^{-\frac{1}{2}\beta_p L} \\ 0 & 0 & \beta_p \, e^{\beta_p L} & -\beta_p \, e^{-\beta_p L} \end{bmatrix}.\tag{4.54}$$

Equation (4.54) has nonzero solutions if

$$|\mathbf{M}| = -\alpha^2\beta_p^3[(\zeta - \alpha)(e^{-\beta_p\gamma_1} - e^{\beta_p\gamma_1}) + (\zeta + \alpha)(e^{-\beta_p\gamma_2} - e^{\beta_p\gamma_2})] = 0,\tag{4.55}$$

where $\gamma_1 = L(\alpha - 1)/2$ and $\gamma_2 = L(\alpha + 1)/2$. Note that the eigenvalue (4.55) reduces to (4.11) if $\alpha = 1$ and $\zeta = 1$ because this corresponds to the single-domain problem with $D_p = D_m$ and $\varepsilon_p = \varepsilon_m$. Equation (4.55) can also be written using hyperbolic functions as

$$\beta_p^3[(\zeta - \alpha)\sinh(\beta_p\gamma_1) + (\zeta + \alpha)\sinh(\beta_p\gamma_2)] = 0.\tag{4.56}$$

Equation (4.55) has only imaginary roots $\beta_p = \sqrt{\varepsilon_p\lambda/D_p}$ corresponding to real and negative eigenvalues $\lambda < 0$. These roots are found numerically and substituted into Equation (4.54), making $\mathbf{M}$ singular. The eigenvector corresponding to the zero eigenvalue provides the eigenfunction coefficients $C_{1m}, \ldots, C_{2p}$. The eigenfunctions are again orthogonal, so if we substitute the eigenfunction expansion (4.17) into the field equations (4.41) and (4.42) and integrate over the domain we will obtain the same state matrices as in Equation (4.21). The elements of the $\mathbf{B}$ vector, however, become

$$b_n = \int_0^{L/2} bC_n(x)\,dx - \int_{L/2}^L bC_n(x)\,dx.\tag{4.57}$$

In this example, we take the output $y(t) = c(L, t) - c(0, t)$ because the output voltage for a cell typically depends on the concentration difference between the two electrodes. The output is expressed as an eigenfunction series evaluated at $x = L$ minus $x = 0$ as in Equation (4.24).

Transfer Function
Laplace transform of Equations (4.41) and (4.42) produces

$$s\varepsilon_m C_m - D_m C_m'' - bI = 0 \quad \text{for } x \in (0, L/2),\tag{4.58}$$

$$s\varepsilon_p C_p - D_p C_p'' + bI = 0 \quad \text{for } x \in (L/2, L).\tag{4.59}$$

The solutions of Equations (4.58) and (4.59) are

$$C_m(x) = C_{1m}\, e^{\beta_m x} + C_{2m}\, e^{-\beta_m x} + \frac{bI}{\varepsilon_m s}, \tag{4.60}$$

$$C_p(x) = C_{1p}\, e^{\beta_p x} + C_{2p}\, e^{-\beta_p x} - \frac{bI}{\varepsilon_p s}. \tag{4.61}$$

Substitution of Equations (4.60) and (4.61) into Equations (4.58) and (4.59) produces

$$s = \frac{D_m \beta_m^2}{\varepsilon_m} = \frac{D_p \beta_p^2}{\varepsilon_p} \tag{4.62}$$

and the same relationships in Equations (4.51) and (4.52). Substitution of Equation (4.62) into the solutions (4.60) and (4.61) and then into the boundary conditions (4.44) and (4.45) produces four linear equations in four unknowns C_{1m}, C_{2m}, C_{1p}, and C_{2p}. The transfer function

$$\frac{D_p Y(s)}{b\varepsilon_p I(s)} = \frac{4\alpha \sinh(\frac{1}{2}\beta_p L) - 2(\zeta - \alpha)\sinh(\beta_p \gamma_1)}{+4\zeta \sinh(\frac{1}{2}\alpha\beta_p L) - 2(\zeta + \alpha)\sinh(\beta_p \gamma_2)}{\beta_p^2[(\zeta - \alpha)\sinh(\beta_p \gamma_1) + (\zeta + \alpha)\sinh(\beta_p \gamma_2)]} \tag{4.63}$$

results from substituting these solutions into $Y(s) = C(L, s) - C(0, s)$, where $\beta_p = \sqrt{\varepsilon_p s / D_p}$. The characteristic equation corresponding to the denominator of Equation (4.63) matches that calculated from the eigenvalue approach in Equation (4.56).

4.1.2 Coupled Electrolyte–Solid Diffusion in Pb Electrodes

In this section we study the coupled concentration and potential governing equations of a Pb–acid battery. To simplify the analysis, only the negative (Pb) electrode is modeled. The equations can be cast in the following form:

$$\left.\begin{aligned} \dot{c} &= a_1 c'' - a_2 \eta \\[2mm] \eta'' &= a_3 \eta + a_4 c'' \end{aligned}\right\} \quad \text{for } x \in (0, L) \tag{4.64}$$

with boundary conditions

$$c'(0, t) = c'(L, t) = 0, \tag{4.65}$$

$$\eta'(0, t) = -a_5 I \text{ and } \eta'(L, t) = -a_6 I, \tag{4.66}$$

where $c(x, t)$ is the acid concentration and $\eta(x, t)$ is the overpotential. The output voltage

$$V(t) = -\eta(L, t). \tag{4.67}$$

The parameters $a_1, \ldots, a_6$ are positive constants. The equations in (4.64) constitute a set of coupled, linear PDEs. Only the first equation involves time derivatives, so the second equation can be thought of as a constraint. This complicates the exact solution, especially for the generation of modal, time-domain models. Using a transfer-function approach, however, we eliminate the time derivative and produce the two differential equations

$$sC - a_1 C'' + a_2 \mathcal{N} = 0,$$
$$\mathcal{N}'' - a_3 \mathcal{N} - a_4 C'' = 0,$$

(4.68)

where $C(x, s) = \mathcal{L}\{c(x, t)\}$ and $\mathcal{N}(x, s) = \mathcal{L}\{\eta(x, t)\}$. These equations have a total of four spatial derivatives, so we expect four solutions of the form $e^{\beta x}$. Substitution of $C(x, s) = C(x) e^{\beta x}$ and $\mathcal{N}(x, s) = \mathcal{N}(x) e^{\beta x}$ into Equation (4.68) produces

$$\begin{bmatrix} s - a_1 \beta^2 & a_2 \\ -a_4 \beta^2 & \beta^2 - a_3 \end{bmatrix} \begin{bmatrix} C(x) \\ \mathcal{N}(x) \end{bmatrix} = \mathbf{0}.$$

(4.69)

To have nonzero solutions for $C(x)$ and $\mathcal{N}(x)$ the determinant of the matrix in Equation (4.69) must be zero, providing the four roots

$$\beta = \pm \sqrt{\frac{s + a_7 \pm \sqrt{(s + a_7)^2 - 4 a_1 a_3 s}}{2 a_1}},$$

(4.70)

where $a_7 = a_1 a_3 + a_2 a_4$. Taking the plus and minus inside the square root as two distinct roots, the four roots simplify to $\beta = \pm \beta_1, \pm \beta_2$. The eigenvectors of the matrix in Equation (4.69) take the form

$$\mathbf{V}_j = \begin{bmatrix} 1 \\ d_j \end{bmatrix} \quad \text{for } j = 1, 2,$$

(4.71)

where $d_j = (a_1 \beta_j^2 - s)/a_2$. The solutions to (4.68) take the form

$$\left. \begin{array}{l} C(x, s) = c_1 e^{\beta_1 x} + c_2 e^{-\beta_1 x} + c_3 e^{\beta_2 x} + c_4 e^{-\beta_2 x} \\[2mm] \mathcal{N}(x, s) = d_1 (c_1 e^{\beta_1 x} + c_2 e^{-\beta_1 x}) + d_2 (c_3 e^{\beta_2 x} + c_4 e^{-\beta_2 x}) \end{array} \right\} \quad \text{for } x \in (0, L). \quad (4.72)$$

Substitution of (4.72) into the boundary conditions (4.65) and (4.66) yields

$$\begin{bmatrix} \beta_1 & -\beta_1 & \beta_2 & -\beta_2 \\ \beta_1 e^{\beta_1 L} & -\beta_1 e^{-\beta_1 L} & \beta_2 e^{\beta_2 L} & -\beta_2 e^{-\beta_2 L} \\ d_1 \beta_1 & -d_1 \beta_1 & d_2 \beta_2 & -d_2 \beta_2 \\ d_1 \beta_1 e^{\beta_1 L} & -d_1 \beta_1 e^{-\beta_1 L} & d_2 \beta_2 e^{\beta_2 L} & -d_2 \beta_2 e^{-\beta_2 L} \end{bmatrix} \begin{bmatrix} c_1 \\ c_2 \\ c_3 \\ c_4 \end{bmatrix} = \begin{bmatrix} 0 \\ 0 \\ -a_5 \\ -a_6 \end{bmatrix} I. \quad (4.73)$$

To calculate the frequency response, we substitute $s = i\omega$ into Equation (4.70) to calculate β_1 and β_2. These are substituted into Equation (4.73) and the matrix is inverted and multiplied with the right-hand side to solve for $c_1, \ldots, c_4$. The coefficients are then substituted into

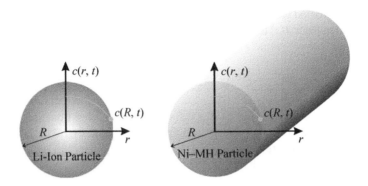

Figure 4.3 Solid-state diffusion for Li-ion (left) and Ni–MH (right) particles

(4.72) to calculate the concentration and overpotential distributions as a function of frequency. The output is calculated from the overpotential distribution evaluated at $x = L$ according to Equation (4.67).

4.1.3 Solid-State Diffusion in Li-Ion and Ni–MH Particles

Figure 4.3 shows solid-state diffusion problems for a spherical Li-ion particle and a cylindrical Ni–MH particle. Conservation of species in the material particle is expressed as

$$\frac{\partial c}{\partial t} = D \left(\frac{\partial^2 c}{\partial r^2} + \frac{m}{r} \frac{\partial c}{\partial r} \right) \quad \text{for } r \in (0, R), \tag{4.74}$$

where $r \in (0, R)$ is the radial coordinate, $c(r, t)$ is the concentration of Li or hydroxide ions in the particle as a function of radial position and time, and D is the solid-phase diffusion coefficient. For the spherical Li-ion particles, $m = 2$. For the cylindrical Ni–MH particles, we assume an infinite length (neglecting the effect of diffusion through the ends of the cylinder) and use $m = 1$. The boundary conditions are

$$\left. \frac{\partial c}{\partial r} \right|_{r=0} = 0, \tag{4.75}$$

$$\left. \frac{\partial c}{\partial r} \right|_{r=R} = \frac{-j}{D a_s F}, \tag{4.76}$$

where $j(t)$ is the volumetric rate of electrochemical reaction at the particle surface (with $j > 0$ indicating ion discharge), a_s is the specific interfacial surface area, and F is the Faraday constant (96 487 C/mol). For the spherical Li-ion and cylindrical Ni–MH particles occupying electrode volume fraction ε, the interfacial surface areas are $a_s = 3\varepsilon/R$ and $2\varepsilon/R$, respectively.

For solid-state diffusion we are most interested on the surface concentration of the particles so a transfer function from $J(s)$ to $C(R, s)$ is sought. Taking the Laplace transform of Equation (4.74) yields

$$C'' + \frac{m}{r}C' - \frac{s}{D}C = 0. \tag{4.77}$$

Finding the transfer function $G(s) = C(R, s)/J(s)$ involves solving Equation (4.77) subject to the Laplace transforms of the boundary conditions (4.75) and (4.76). It is convenient to define $\Gamma(s) = R\sqrt{s/D}$ in the following. For a detailed solution, see [19].

For the cylindrical particles with $m = 1$, the transfer function is

$$G(s) = -\frac{R\, I_0(\Gamma(s))}{Da_s F[\Gamma(s)I_1(\Gamma(s))]}, \tag{4.78}$$

where $I_0(x)$ and $I_1(x)$ are zeroth- and first-order modified Bessel functions of the first kind.

For spherical particles with $m = 2$, the transfer function is

$$G(s) = \frac{\tanh(\Gamma(s))}{Da_s F[\tanh(\Gamma(s)) - \Gamma(s)]}. \tag{4.79}$$

4.2 Padé Approximation Method

The analytical solutions for battery-cell-related models can often be expressed in terms of a transcendental transfer function like the spherical and cylindrical particles in the previous section. These transfer functions often involve hyperbolic functions that can also be written in terms of exponentials. The Padé approximation works well for these infinitely differentiable functions that can be expanded in a power series at the origin. The Nth-order Padé approximation of a transfer function $G(s)$ is a ratio of two polynomials in s where the denominator is of order N. For a proper transfer function, the numerator is of order N or less. The Padé approximation method can produce transfer functions with numerators of order 1 to N, but the highest accuracy is often obtained when the numerator order is N. One may also consider the high-frequency response to determine the proper order of the numerator. The computational speed of the model and any model-based estimators and controllers depend strongly on the number of integrators in the model or the order of the denominator. The order of the numerator, however, does not influence computational speed as significantly.

Following [20], we assume that the transfer function can be expanded in a power series at the origin as follows

$$G(s) = \sum_{k=0}^{2(N+1)} c_k s^k, \tag{4.80}$$

where the coefficients c_k are calculated by repeated differentiation and evaluation of $G(s)$ at $s = 0$,

$$c_k = \left.\frac{d^k G(s)}{ds^k}\right|_{s=0}. \tag{4.81}$$

If $G(s)$ has a pole at the origin then we apply the power series expansion to $G^*(s) = sG(s)$.
The Nth-order Padé approximation transfer function

$$P(s) = \frac{\sum_{m=0}^{N} b_m s^m}{1 + \sum_{n=1}^{N} a_n s^n} = \frac{\text{num}(s)}{\text{den}(s)}, \tag{4.82}$$

where we assume that the denominator and numerator both have order N. To determine $P(s)$ we must calculate the $(N + 1)$ b_m and N a_m coefficients. The zeroth-order term in the denominator is assumed to have a unity coefficient to normalize the solutions. The $2N + 1$ linear equations that can be solved for the coefficients are determined from the polynomial equation

$$\text{den}(s) \sum_{k=0}^{2(N+1)} c_k s^k - \text{num}(s) = 0, \tag{4.83}$$

where the coefficients c_k are known from the power series expansion. Equation (4.83) produces a polynomial of order $2N(N + 1)$ in s. The right-hand side equals zero for all s, so the coefficients must be zero. The first $N + 1$ coefficients of s depend on both the unknown a_n and b_n coefficients. The remaining coefficients depend only on a_n. Thus, we set the coefficients of s^{N+2} to s^{2N+1} equal to zero to solve for $a_1, \ldots, a_N$. Then we substitute these solutions $a_1, \ldots, a_N$ into the coefficients of s^0 to s^N and set them equal to zero to solve for $b_0, \ldots, b_N$.

4.2.1 Solid-State Diffusion in Li-Ion Particles

The transfer function in Equation (4.79) can be written in the form

$$G(s) = \frac{\sinh(\sqrt{s^*})}{\sinh(\sqrt{s^*}) - \sqrt{s^*} \cosh(\sqrt{s^*})} \tag{4.84}$$

where we have substituted $s = Ds^*/R^2$ and normalized by DAF. The transfer function (4.84) has a power series

$$sG(s) = -3 - \frac{1}{5}s + \frac{1}{175}s^2 - \frac{2}{7875}s^3 + \frac{37}{3\,031\,875}s^4 + \cdots, \tag{4.85}$$

where premultiplication by s cancels the singularity at $s = 0$ and the stars have been dropped for simplicity. For $N = 2$, Equation (4.83) becomes

$$\frac{37a_2}{3\,031\,875}s^6 + \left(\frac{37a_1}{3\,031\,875} - \frac{2a_2}{7875}\right)s^5 + \left(\frac{37}{3\,031\,875} - \frac{2a_1}{7875} + \frac{a_2}{175}\right)s^4$$

$$+ \left(\frac{a_1}{175} - \frac{2}{7875} - \frac{a_2}{5}\right)s^3 + \left(\frac{1}{175} - b_2 - \frac{a_1}{5} - 3a_2\right)s^2$$

$$+ \left(-3a_1 - \frac{1}{5} - b_1\right)s - 3 - b_0 = 0. \tag{4.86}$$

The s^4 and s^3 coefficients provide two equations in the two unknowns a_1 and a_2. Then, the s^2, s, and constant terms provide equations for b_0, b_1, and b_2. Substituting the solutions into (4.82) produces the second-order Padé approximation

$$s P(s) = \frac{-3 - \frac{4}{11}s - \frac{1}{165}s^2}{1 + \frac{3}{55}s + \frac{1}{3465}s^2}. \tag{4.87}$$

In this example, the Padé approximation is actually order $N + 1$ owing to the extra s in the denominator.

4.3 Integral Method Approximation

Another way to convert the governing PDEs of a battery model to ODEs is to assume a distribution across the cell for the distributed variable of interest and integrate the governing equations to convert the PDE to a single ODE. In this section we apply the IMA to planar, cylindrical, and spherical diffusion. More information on the IMA can be found in [18, 21–23].

4.3.1 Electrolyte Diffusion

Single Domain

To demonstrate the IMA approach, we first apply the IMA to the single-domain electrolyte diffusion problem described in Section 4.1.1. If we assume that the concentration has a uniform distribution $c(x, t) = c_0(t)$, then the zero-flux boundary conditions (4.4) are satisfied. Substituting into Equation (4.3) yields

$$\dot{c}_0(t) = a_2 j(t). \tag{4.88}$$

The ODE (4.88) provides a lumped-parameter approximation of the concentration dynamics in the electrolyte. In this case, however, the lumped-parameter solution is also the exact solution.

For the concentration flux boundary condition (4.5), a uniform concentration distribution cannot satisfy the boundary conditions, so we assume a parabolic distribution:

$$c(x, t) = c_0(t) + c_1(t)x + c_2(t)x^2. \tag{4.89}$$

The three unknown functions in Equation (4.89) are solved for from the field equation (4.3) and two boundary conditions. Differentiation of Equation (4.89) yields

$$\frac{\partial c}{\partial x} = c_1(t) + 2c_2(t)x. \tag{4.90}$$

Evaluation of Equation (4.90) at $x = 0$ using boundary condition (4.4) gives $c_1(t) = 0$. Substitution of the assumed distribution (4.89) into boundary condition (4.5) results in

$$2c_2(t)L = a_3 I(t) \quad \text{or} \quad c_2(t) = \frac{a_3}{2L} I(t). \tag{4.91}$$

We substitute Equation (4.91) into Equation (4.89) and then into Equation (4.3) with $j = 0$ and then integrate from $x = 0 \dots L$ to obtain

$$\dot{c}_0(t) = -\frac{a_3 L}{6} \dot{I}(t) + \frac{a_1 a_3}{L} I(t). \tag{4.92}$$

Time integration of Equation (4.92) and

$$c(x, t) = c_0(t) + \frac{a_3}{2L} x^2 I(t) \tag{4.93}$$

gives the concentration distribution as a function of time.

Equations (4.92) and (4.93) can be put in state-variable form (4.1) by defining the state vector $\mathbf{x}(t) = c_0(t) + \frac{a_3 L}{6} I(t)$ and output $y(t) = c(L, t)$. The input $u(t) = I(t)$. The state matrices then become

$$A = 0, \quad B = \frac{a_1 a_3}{L}, \quad C = 1, \quad D = \frac{a_3 L}{3}. \tag{4.94}$$

We can generate the transfer function $G(s) = c(L, s)/I(s)$ by substituting the Laplace transform of Equation (4.92) into Equation (4.93) evaluated at $x = L$ to produce

$$G(s) = \frac{a_3(L^2 s + 3a_1)}{3Ls}. \tag{4.95}$$

We can extend the IMA to higher order polynomials, and presumably higher accuracy, by adding one additional equation for each additional coefficient in the polynomial. One way to generate additional equations is to evaluate the field equation (4.3) at additional points (e.g., $x = L/4, L/2, 3L/4, \dots$) in what is called a collocation method. Section 4.4, however, presents a more efficient method of extending the IMA to higher order polynomials called the Ritz method.

Coupled Domains

Referring back to the electrolyte diffusion problem with coupled domains described in Section 4.1.1, the IMA assumes that the concentration has parabolic distributions in each domain:

$$c(x, t) = \begin{cases} c_{0m}(t) + c_{1m}(t)x + c_{2m}(t)x^2 & \text{for } x \leq L/2, \\ c_{0p}(t) + c_{1p}(t)x + c_{2p}(t)x^2 & \text{for } x \geq L/2. \end{cases} \tag{4.96}$$

The six coefficients in Equation (4.96) can be solved from the two field equations ($x < L/2$ and $x > L/2$) and the four boundary conditions (4.44) and (4.45). Substitution of (4.96) into the Laplace transform of Equation (4.3) and integration yields

$$\int_0^{L/2} (s\varepsilon_m C - D_m C'' - bI) \, dx$$

$$= \frac{\varepsilon_m L}{2} s C_{0m} + \frac{\varepsilon_m L^2}{8} s C_{1m} + \left(\frac{\varepsilon_m L^3 s}{24} - LD_m \right) C_{2m} - \frac{bL}{2} I = 0 \tag{4.97}$$

and

$$\int_{L/2}^{L} (s\varepsilon_p C - D_p C'' + bI)\, dx$$

$$= \frac{\varepsilon_p L}{2} s C_{0p} + \frac{3\varepsilon_p L^2}{8} s C_{1p} + \left(\frac{7\varepsilon_p L^3}{24} s - LD_p \right) C_{2p} + \frac{bL}{2} I = 0. \qquad (4.98)$$

Equation (4.44) gives

$$C_{1m} = 0 \quad \text{and} \quad C_{1p} + 2LC_{2p} = 0. \qquad (4.99)$$

Substitution of Equations (4.96) and (4.99) into boundary conditions (4.45) yields

$$D_m C_{2m} + D_p C_{2p} = 0$$

$$C_{0m} + \frac{L^2}{4} C_{2m} - C_{0p} + \frac{3L^2}{4} C_{2p} = 0 \qquad (4.100)$$

Solution of Equations (4.97)–(4.100) and substitution into the output equation

$$Y(s) = C(L, s) - C(0, s) = C_{0p} + C_{1p} L + C_{2p} L^2 - C_{0m} \qquad (4.101)$$

gives the transfer function

$$G(s) = \frac{Y(s)}{I(s)} = \frac{-3\, b\, L^2 (\varepsilon_m + \varepsilon_p)(D_m + D_p)}{2\varepsilon_m \varepsilon_p L^2 (D_m + D_p)s + 24 D_m D_p (\varepsilon_m + \varepsilon_p)}. \qquad (4.102)$$

The IMA can also be applied to two or more coupled potential equations within a single planar domain. The equations can include time derivatives or only spatial derivatives. For second-order (diffusion-type) differential equations with two spatial derivatives, a parabolic distribution provides sufficient unknown coefficients to satisfy the volume-averaged field equation and the two boundary conditions. If the field equation does not include time derivatives, then the resulting equations will be algebraic. If the governing equations are linear, then the integral method approach always produces a combination of linear algebraic and differential equations that must be solved simultaneously.

The IMA can also be extended to higher order approximations by evaluating the field equation at specific points in the domain. For each additional term in the approximation, an additional equation is added. In the coupled-domain problem, for example, we can add $c_{3m} x^3$ to the approximation in Equation (4.96) and solve the additional equation

$$\varepsilon_m \frac{\partial c}{\partial t} \bigg|_{x^*} - D_m \frac{\partial^2 c}{\partial x^2} \bigg|_{x^*} - bI$$

$$= \varepsilon_m (\dot{c}_{0m} + \dot{c}_{2m} x^{*2} + \dot{c}_{3m} x^{*3}) - D_m (2c_{2m} + 3c_{3m} x^*) - bI = 0, \qquad (4.103)$$

where $x^* \in [0, L/2]$. Equation (4.103) is a first-order differential equation, increasing the order of the approximation by one. Additional terms can be added to the approximation in Equation (4.96) with additional equations from Equation (4.103) evaluated at different x^*.

4.3.2   Solid-State Diffusion in Li-Ion and Ni–MH Particles

For the spherical Li-ion particles ($m = 2$) and cylindrical Ni–MH particles ($m = 1$) described in Section 4.1.3 we use a quadratic approximation across the radius of the particle as follows:

$$C(r, s) = C_0(s) + C_1(s)r + C_2(s)r^2. \tag{4.104}$$

Zero flux at the origin requires $C_1(s) = 0$. The boundary flux at the surface of the particle requires

$$C'(R, s) = 2C_2 R = -\frac{J(s)}{DAF}. \tag{4.105}$$

The final equation comes from integration of the Laplace-transformed field equation (4.74) to produce

$$\int_0^R \left[\frac{s}{D}C - \left(C'' + \frac{m}{r}C' \right) \right] dr = 0. \tag{4.106}$$

Substitution of Equation (4.104) into Equation (4.106) yields

$$C_0(s) = -\frac{[6D(1 + m) - sR^2]D}{6RDAFs} J(s). \tag{4.107}$$

Substitution of $C_0(s)$, $C_1(s)$, and $C_2(s)$ into Equation (4.104) and evaluation at $r = R$ produces the transfer function

$$G(s) = \frac{C(R, s)}{J(s)} = -\frac{R^2 s + 3D(m + 1)}{3RDAFs}. \tag{4.108}$$

4.4   Ritz Method

The Ritz method maintains the inherent symmetry of the operators in the governing PDEs. In battery systems, diffusion equations are symmetric, producing real eigenvalues and an exponentially decaying response. The discretized $\mathbf{A}$ matrices generated by the Ritz method are also symmetric, ensuring real eigenvalues. The convergence properties of Ritz expansions have also been thoroughly studied. The eigenvalue magnitudes converge monotonically from above as the order of the series increases.

4.4.1   Electrolyte Diffusion in a Single Domain

We first apply the Ritz method to the single-domain problem described in Section 4.1.1. The distributed function $c(x, t)$ is approximated by the polynomial series

$$c(x, t) = \sum_{n=0}^{N-1} x^n c_n(t). \tag{4.109}$$

Equation (4.109) does not automatically satisfy the boundary conditions. Converting Equation (4.3) to a weak form automatically accounts for flux boundary conditions that only involve partial derivatives of the concentration. The weak form of Equation (4.3) is obtained by premultiplying Equation (4.3) by a polynomial and integrating to generate

$$\int_0^L x^n \frac{\partial c}{\partial t}\, dx = \int_0^L x^n \left([a_1 \frac{\partial^2 c}{\partial x^2} + a_2 j \right) dx \tag{4.110}$$

$$= a_1 x^n \frac{\partial c}{\partial x}\Big|_0^L + \int_0^L \left(a_2 j x^n - a_1 n x^{n-1} \frac{\partial c}{\partial x} \right) dx \tag{4.111}$$

$$= a_1 a_3 L^n I(t) + \frac{a_2 L^{n+1}}{n+1} j(t) - a_1 n \int_0^L \left(x^{n-1} \frac{\partial c}{\partial x} \right) dx. \tag{4.112}$$

In Equation (4.111), we have used integration by parts and then substituted the boundary conditions to obtain Equation (4.112). Substitution of the expansion (4.109) into (4.112) yields

$$\int_0^L \left(x^n \sum_{m-0}^{N-1} x^m \dot{c}_m(t) \right) dx = a_1 a_3 L^n I(t) + \frac{a_2 L^{n+1}}{n+1} j(t)$$

$$- a_1 n \int_0^L \left(x^{n-1} \sum_{m=0}^{N-1} m x^{m-1} c_m(t) \right) dx \tag{4.113}$$

Exchanging summation and integration, we obtain the discretized equations

$$\sum_{m=0}^{N-1} \left(\int_0^L x^{n+m}\, dx\, \dot{c}_m(t) \right) = a_1 a_3 L^n I(t) + \frac{a_2 L^{n+1}}{n+1} j(t)$$

$$- a_1 n \sum_{m=0}^{N-1} \left(m \int_0^L x^{n+m-2}\, dx\, c_m(t) \right) \tag{4.114}$$

$$\text{for } n = 0, \ldots, N.$$

Equation (4.114) can be put in state-variable form (4.1), where $\mathbf{x}(t) = [c_0, \ldots, c_{N-1}]^T$, $\mathbf{u}(t) = [I(t), j(t)]^T$,

$$\mathbf{A} = \mathbf{M}_1^{-1}\mathbf{M}_2, \quad \mathbf{B} = \mathbf{M}_1^{-1}\mathbf{M}_3,$$
$$\mathbf{C} = c(L, t) = [1, L, \ldots, L^{N-1}], \quad \text{and } \mathbf{D} = 0, \tag{4.115}$$

where

$$\mathbf{M}_1(k, l) = \int_0^L x^{n+m} \, dx = \frac{L^{n+m+1}}{n+m+1} = \mathbf{M}_1(l, k), \tag{4.116}$$

$$\mathbf{M}_2(k, l) = a_1 nm \int_0^L x^{n+m-2} \, dx = a_1 nm \frac{L^{n+m-1}}{n+m-1} = \mathbf{M}_2(l, k), \tag{4.117}$$

$$\mathbf{M}_3 = \begin{bmatrix} a_1 a_3 & a_2 L \\ \vdots & \vdots \\ a_1 a_3 L^N & \frac{a_2 L^{N+1}}{N+1} \end{bmatrix}, \tag{4.118}$$

where $k = n+1$ and $l = m+1$ are the matrix indices (must be greater than zero). The symmetry of $\mathbf{M}_1$ and $\mathbf{M}_2$ ensures that the discretized linear system has real eigenvalues.

4.4.2 Electrolyte Diffusion in Coupled Domains

The coupled planar-domains model problem described in Section 4.1.1 consists of the field equation

$$\varepsilon(x)\frac{\partial c}{\partial t} = D(x)\frac{\partial^2 c}{\partial x^2} + b(x)I \quad \text{for } x \in (0, L), \tag{4.119}$$

where

$$\varepsilon(x) = \begin{cases} \varepsilon_m & \text{for } 0 < x < L/2 \\ \varepsilon_p & \text{for } L/2 < x < L \end{cases}, \quad D(x) = \begin{cases} D_m & \text{for } 0 < x < L/2 \\ D_p & \text{for } L/2 < x < L \end{cases}, \tag{4.120}$$

and

$$b(x) = \begin{cases} b & \text{for } 0 < x < L/2 \\ -b & \text{for } L/2 < x < L \end{cases}, \tag{4.121}$$

with boundary conditions (4.44) and (4.45). The response will be approximated by admissible functions that are continuous across the domain $0 \le x \le L$. First, the polynomial series in Equation (4.109) is used. Then we use a Fourier series solution with functions that satisfy the zero-flux boundary conditions at $x = 0, L$. Both cases start with a weak form of the governing equation (4.119) that incorporates the natural (flux) boundary conditions. The requirement for continuous concentration at the interface of the two domains will be automatically satisfied by the continuity of the polynomial and sinusoidal functions.

The field equation (4.119) is converted to a weak form by premultiplying by an admissible function $C_n(x)$ and integrating,

$$\int_0^L \varepsilon(x) C_n \dot{c} \, dx = \int_0^L C_n [D(x) c'' + b(x) I] \, dx$$

$$= D_m C_n c' \big|_0^{L/2} + D_p C_n c' \big|_{L/2}^L + \int_0^L C_n b I - D C_n' c' \, dx$$

$$= -\int_0^L D C_n' c' \, dx + b \left(\int_0^{L/2} C_n \, dx - \int_{L/2}^L C_n \, dx \right) I, \quad (4.122)$$

using integration by parts and the boundary conditions. The integral on the right-hand side of Equation (4.122) is broken into two integrations from $x = 0$ to $x = L/2$ and from $x = L/2$ to $x = L$. Substitution of the Ritz expansion

$$c(x, t) = \sum_{m=0}^{N-1} C_m(x) c_m(t) \quad (4.123)$$

into Equation (4.122) produces

$$\int_0^L C_n \sum_{m=0}^{N-1} C_m \varepsilon \dot{c}_m \, dx + \int_0^L D C_n' \sum_{m=0}^{N-1} C_m' c_m \, dx - \left(\int_0^L b C_n \, dx \right) I$$

$$= \sum_{m=0}^{N-1} \left[\left(\int_0^L \varepsilon C_n C_m \, dx \right) \dot{c}_m + \left(\int_0^L D C_n' C_m' \, dx \right) c_m \right]$$

$$- b \left(\int_0^{L/2} C_n \, dx - \int_{L/2}^L C_n \, dx \right) I = 0 \quad (4.124)$$

or

$$\mathbf{M}_1 \dot{\mathbf{x}} = \mathbf{M}_2 \mathbf{x} + \mathbf{M}_3 I, \quad (4.125)$$

where $\mathbf{x}(t) = [c_0(t), \ldots, c_{N-1}(t)]^T$ and

$$\mathbf{M}_1(k, l) = \varepsilon_m \int_0^{L/2} C_n C_m \, dx + \varepsilon_p \int_{L/2}^L C_n C_m \, dx = \mathbf{M}_1(l, k),$$

$$\mathbf{M}_2(k, l) = -D_m \int_0^{L/2} C_n' C_m' \, dx - D_p \int_{L/2}^L C_n' C_m' \, dx = \mathbf{M}_2(l, k),$$

$$\mathbf{M}_3(k) = b \left(\int_0^{L/2} C_n \, dx - \int_{L/2}^L C_n \, dx \right).$$

The discretized differential equations in Equation (4.123) can then be written in state-space form with

$$\mathbf{A} = \mathbf{M}_1^{-1} \mathbf{M}_2, \quad \mathbf{B} = \mathbf{M}_1^{-1} \mathbf{M}_3,$$
$$\mathbf{C} = c(L, t) - c(0, t) = [C_0(L) - C_0(0), \ldots, C_{N-1}(L) - C_{N-1}(0)], \quad (4.126)$$

and $\mathbf{D} = 0$. Again, we obtain symmetric $\mathbf{M}_1$ and $\mathbf{M}_2$ matrices, so the eigenvalues are real.

Polynomial Series

Substitution of polynomial functions $C_n(x) = x^n$ into Equation (4.123) produces

$$\mathbf{M}_1(k, l) = \frac{\varepsilon_m + \varepsilon_p(2^{n+m+1} - 1)}{n + m + 1} \left(\frac{L}{2}\right)^{n+m+1},$$

$$\mathbf{M}_2(1, l) = 0,$$

$$\mathbf{M}_2(k, l) = -\frac{nm[D_m + D_p(2^{n-1+m} - 1)]}{n - 1 + m} \left(\frac{L}{2}\right)^{n-1+m} \qquad \text{for } n, m > 0,$$

$$\mathbf{M}_3(k) = \frac{bL[(L/2)^n - L^n]}{n + 1}, \tag{4.127}$$

with the output vector $\mathbf{C} = [0, L, \ldots, L^{N-1}]$. The first state is uncontrollable because $\mathbf{M}_3(1) = 0$ and $\mathbf{M}_2(1, l) = 0$ and unobservable because $\mathbf{C}(1) = 0$.

Fourier Series

In a Fourier series approximate solution the admissible functions are sinusoidal with $C_n(x) = \cos(n\pi x/L)$. Unlike the polynomial functions, these functions automatically satisfy the zero-flux boundary conditions at $x = 0$ and $x = L$, as shown in Figure 4.4. Substitution into (4.127) produces

$$\mathbf{M}_1(1, 1) = \frac{L}{2}(\varepsilon_m + \varepsilon_p), \tag{4.128}$$

$$\mathbf{M}_1(k, k) = \frac{L}{4}(\varepsilon_m + \varepsilon_p),$$

$$\mathbf{M}_1(k, l) = 0 \quad \text{if } n \text{ and } m \text{ are both odd or both even and } n \neq m,$$

$$\mathbf{M}_1(k, l) = \frac{nL(\varepsilon_m - \varepsilon_p)(-1)^{(n+m-1)/2}}{\pi(n^2 - m^2)} \quad \text{if } n \text{ is odd and } m \text{ is even,}$$

$$\mathbf{M}_2(k, k) = -\frac{n^2\pi^2\left(D_m + D_p\right)}{4L}, \tag{4.129}$$

$$\mathbf{M}_2(k, l) = 0 \quad \text{if } n \text{ and } m \text{ are both odd or both even and } n \neq m,$$

$$\mathbf{M}_2(k, l) = \frac{\pi nm^2(D_m - D_p)(-1)^{(n+m-1)/2}}{L(m^2 - n^2)} \quad \text{if } n \text{ is odd and } m \text{ is even,}$$

$$\mathbf{M}_3(k) = 0 \quad \text{if } n \text{ is even,} \tag{4.130}$$

$$\mathbf{M}_3(k) = \frac{2bL(-1)^{(n-1)/2}}{n\pi} \quad \text{if } n \text{ is odd.}$$

Again, we use $k = n + 1$ and $l = m + 1$ to keep the matrix indices positive. The output matrix $\mathbf{C} = [0, -2, 0, -2, \ldots]$, indicating that the even modes are not directly observable in the output. From Equation (4.130), the even modes are also not directly controllable from the input. The odd modes are coupled to the even modes through the nondiagonal terms in $\mathbf{M}_2$

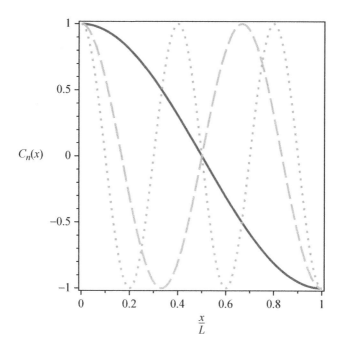

Figure 4.4   Fourier series admissible functions: $n = 1$ (solid), $n = 3$ (dashed), and $n = 5$ (dotted)

and M_1, so they influence the system response. A different output (e.g., $c(L/2, t) - c(0, t)$) would directly sense the even modes.

4.4.3   Coupled Electrolyte–Solid Diffusion in Pb Electrodes

Referring back to the coupled concentration and potential governing equations of a Pb–acid battery introduced in Section 4.1.2, the Ritz method provides a relatively easy way to generate a time-domain model in state-variable form. The equations are restated here as

$$\dot{c} = a_1 c'' - a_2 \eta \tag{4.131}$$

$$\eta'' = a_3 \eta + a_4 c'' \tag{4.132}$$

for $x \in (0, L)$ with boundary conditions

$$c'(0, t) = c'(L, t) = 0, \tag{4.133}$$

$$\eta'(0, t) = -a_5 I \quad \text{and} \quad \eta'(L, t) = -a_6 I, \tag{4.134}$$

where $c(x, t)$ is the acid concentration and $\eta(x, t)$ is the overpotential. The output voltage

$$V(t) = -\eta(L, t). \tag{4.135}$$

The concentration and overpotential are expanded in Fourier series:

$$c(x, t) = \sum_{n=0}^{N-1} c_n(t) \cos\left(\frac{n\pi x}{L}\right),$$

$$\eta(x, t) = \sum_{n=0}^{N-1} \eta_n(t) \cos\left(\frac{n\pi x}{L}\right). \tag{4.136}$$

The cosine expansions in (4.136) have zero slope at $x = 0, L$, matching the zero input boundary conditions. The equations in (4.136) are substituted into Equation (4.131), premultiplied by $\cos(m\pi x/L)$ and integrated over the domain to produce

$$\mathbf{M}\dot{\mathbf{c}} = -\mathbf{K}_{cc}\mathbf{c} - \mathbf{K}_{c\eta}\eta, \tag{4.137}$$

where $\mathbf{c}(t) = [c_1(t), \ldots, c_N(t)]$ and the matrix elements

$$\mathbf{M}(n, m) = \int_0^L \cos\left(\frac{m\pi x}{L}\right) \cos\left(\frac{n\pi x}{L}\right) dx = \frac{L}{2}\delta_{mn},$$

$$\mathbf{K}_{cc}(n, m) = \int_0^L a_1 \left(\frac{n\pi}{L}\right)^2 \cos\left(\frac{m\pi x}{L}\right) \cos\left(\frac{n\pi x}{L}\right) dx = \frac{a_1 n^2 \pi^2}{2L}\delta_{mn},$$

$$\mathbf{K}_{c\eta}(n, m) = \int_0^L a_2 \cos\left(\frac{m\pi x}{L}\right) \cos\left(\frac{n\pi x}{L}\right) dx = \frac{a_2 L}{2}\delta_{mn},$$

where $\delta_{mn} = 1$ if $m = n$ and zero otherwise. For Equation (4.132), we premultiply by $v(x)$ and integrate by parts to produce the weak form

$$(a_5 v(0) - a_6 v(L))I(t) - \int_0^L v'\eta' + a_3 v\eta - a_4 v'c' \, dx = 0. \tag{4.138}$$

Substitute $v(x, t) = \cos[(m\pi x)/L]$ and the Fourier series (4.136) into Equation (4.138) to obtain

$$\mathbf{K}_{\eta\eta}\eta = \mathbf{K}_{\eta c}\mathbf{c} - \mathbf{B}_{\eta}I, \tag{4.139}$$

where the matrix elements

$$\mathbf{K}_{\eta\eta}(m, n) = \int_0^L \frac{m\pi}{L}\sin\left(\frac{m\pi x}{L}\right)\frac{n\pi}{L}\sin\left(\frac{n\pi x}{L}\right) + a_3 \cos\left(\frac{m\pi x}{L}\right)\cos\left(\frac{n\pi x}{L}\right) dx$$

$$= \left[\frac{n^2\pi^2}{2L} + \frac{a_3 L}{2}\right]\delta_{mn},$$

$$\mathbf{K}_{\eta c}(m, n) = \int_0^L a_4 \frac{m\pi}{L}\sin\left(\frac{m\pi x}{L}\right)\frac{n\pi}{L}\sin\left(\frac{n\pi x}{L}\right) dx = \frac{a_4 n^2\pi^2}{2L}\delta_{mn},$$

$$\mathbf{B}_{\eta}(n) = a_5 - a_6(-1)^n.$$

For $n = m = 0$, the nonzero matrix elements change as follows: $\mathbf{M}(0, 0) = L$, $\mathbf{K}_{c\eta}(0, 0) = a_2 L$, and $\mathbf{K}_{\eta\eta}(0, 0) = a_3 L$.

The matrix $\mathbf{K}_{\eta\eta}$ is nonsingular, so, from Equation (4.139),

$$\eta = \mathbf{K}_{\eta\eta}^{-1}(\mathbf{K}_{\eta c}\mathbf{c} - \mathbf{B}_\eta I). \tag{4.140}$$

Substitution of Equation (4.140) into Equation (4.137) produces the state-variable model

$$\dot{\mathbf{x}} = \mathbf{A}\mathbf{x} + \mathbf{B}I, \tag{4.141}$$

$$V = \mathbf{C}\mathbf{x} + \mathbf{D}I, \tag{4.142}$$

where

$$\mathbf{A} = -\mathbf{M}^{-1}[\mathbf{K}_{cc} + \mathbf{K}_{c\eta}\mathbf{K}_{\eta\eta}^{-1}\mathbf{K}_{\eta c}],$$

$$\mathbf{B} = \mathbf{M}^{-1}\mathbf{K}_{c\eta}\mathbf{K}_{\eta\eta}^{-1}\mathbf{B}_\eta,$$

$$\mathbf{C} = \mathbf{v}^{\mathrm{T}}\mathbf{K}_{\eta\eta}^{-1}\mathbf{K}_{\eta c},$$

$$\mathbf{D} = \mathbf{v}^{\mathrm{T}}\mathbf{K}_{\eta\eta}^{-1}\mathbf{B}_\eta,$$

where $\mathbf{v} = [1, \cos(\pi x/L), \ldots, \cos[(N-1)\pi x/L]]$.

4.5 Finite-Element Method

The FEM is based on a weak form of the governing equation as was used in the Ritz method. Rather than choosing functions that exist over the entire domain, however, FEM discretizes the domain $x \in [0, L]$ into $N - 1$ subdomains or elements:

$$\Omega_1 = [0, h], \ldots, \Omega_m = [(m-1)h, mh], \ldots, \Omega_{N-1} = [L - h, L]. \tag{4.143}$$

In general, the length of each element can be varied to improve the accuracy in high-flux regions and reduce the number of elements in regions with low gradients. For simplicity, we assume that the grid is uniform with each element having length h; so $L = h(N - 1)$. The concentrations at the endpoints of the domains are referred to as nodes $c_m(t) = c((m-1)h, t)$ for $m = 1, \ldots, N$. The Nth-order FEM approximation has N nodes. FEM generates equations for the nodal dynamics that can be realized in state-variable or transfer-function forms. For more information and details on the FEM method, readers are referred to [17, 24].

As shown in Figure 4.5, the shape functions $\varphi_m(x)$ for $m = 1, \ldots N - 1$ are defined as

$$\varphi_1(x) = \begin{cases} 1 - x/h, & x \in \Omega_1, \\ 0, & x \notin \Omega_1, \end{cases}$$

$$\varphi_m(x) = \begin{cases} [x - (m-2)h]/h, & x \in \Omega_{m-1}, \\ 1 - [x - (m-1)h]/h, & x \in \Omega_m, \\ 0, & x \notin \Omega_m \cup \Omega_{m-1}, \end{cases} \quad 2 \le m \le N - 1, \tag{4.144}$$

$$\varphi_N(x) = \begin{cases} 1 - [x - (N-2)h]/h, & x \in \Omega_{N-1}, \\ 0, & x \notin \Omega_{N-1}. \end{cases}$$

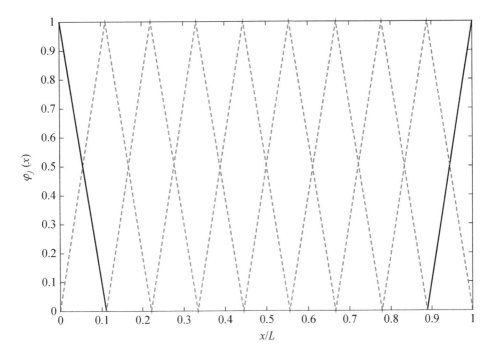

Figure 4.5 FEM admissible functions ($N = 10$)

The shape functions have $\varphi_m[(n-1)h] = \delta_{mn}$, so they are unity at the mth node ($x = (m - 1)h$) and zero at all the other nodes. It also follows that $\varphi_m \varphi_n = 0$ if m and n differ by more than one. The FEM approximation is given by

$$c(x, t) = \sum_{m=1}^{N} c_m(t)\varphi_m(x). \qquad (4.145)$$

4.5.1 Electrolyte Diffusion

For diffusion equations governed by Equation (4.3) and zero-flux boundary conditions (4.4) the weak form arises by integration by parts of the field equation premultiplied by φ_n to obtain

$$\sum_{m=1}^{N}\left(\int_0^L \varphi_n\varphi_m \, dx \, \dot{c}_m + \int_0^L a_1\varphi_n'\varphi_m' \, dx \, c_m\right) - \int_0^L a_2\varphi_n j \, dx = 0, \qquad (4.146)$$

for $n = 1, \ldots, N$. We assume that the coefficients a_1 and a_2 and input j are constant within each element at a_{1_n}, a_{2_n}, and j_n, respectively, but can vary across the domain. The integrals in Equation (4.146) simplify because $\varphi_n = 0$ outside of domains $\Omega_n \cup \Omega_{n-1}$. Thus, the mth equation only involves φ_n that are nonzero in at least one of these domains or c_{m-1}, c_m, and

c_{m+1}. Based on this observation, the integrals in Equation (4.146) simplify as follows:

$$\sum_{m=1}^{N} \int_0^L a_1 \varphi_n' \varphi_m' \, dx \, c_m = a_{1_{n-1}} \int_{\Omega_{n-1}} \varphi_n' \left(\varphi_{n-1}' c_{n-1} + \varphi_n' c_n \right) \, dx$$

$$+ a_{1_n} \int_{\Omega_n} \varphi_n' \left(\varphi_n' c_n + \varphi_{n+1}' c_{n+1} \right) \, dx$$

$$= \left(a_{1_{n-1}} + a_{1_n} \right) \int_0^h \varphi_1'^2 \, dx \, c_n$$

$$+ \int_0^h \varphi_1' \varphi_2' \, dx \left(a_{1_{n-1}} c_{n-1} + a_{1_n} c_{n+1} \right)$$

$$= -\frac{a_{1_{n-1}}}{h} c_{n-1} + \frac{1}{h} \left(a_{1_{n-1}} + a_{1_n} \right) c_n - \frac{a_{1_n}}{h} c_{n+1} \qquad (4.147)$$

using the constant element length assumption. For $n = 1$,

$$\sum_{m=1}^{N} \int_0^L a_1 \varphi_1' \varphi_m' \, dx \, c_m = a_{1_1} \int_{\Omega_1} \varphi_1' \left(\varphi_1' c_1 + \varphi_2' c_2 \right) \, dx$$

$$= \frac{a_{1_1}}{h} c_n - \frac{a_{1_1}}{h} c_2, \qquad (4.148)$$

with a similar result for $n = N$. The first integral in Equation (4.146) becomes

$$\sum_{m=1}^{N} \int_0^L \varphi_n \varphi_m \, dx \, \dot{c}_m = \int_{\Omega_{n-1}} \varphi_n \left(\varphi_{n-1} \dot{c}_{n-1} + \varphi_n \dot{c}_n \right) \, dx$$

$$+ \int_{\Omega_n} \varphi_n \left(\varphi_n \dot{c}_n + \varphi_{n+1} \dot{c}_{n+1} \right) \, dx$$

$$= 2 \int_0^h \varphi_1^2 \, dx \, \dot{c}_n + \int_0^h \varphi_1 \varphi_2 \, dx \left(\dot{c}_{n-1} + \dot{c}_{n+1} \right)$$

$$= \frac{h}{6} \dot{c}_{n-1} + \frac{2h}{3} \dot{c}_n + \frac{h}{6} \dot{c}_{n+1}. \qquad (4.149)$$

The expressions of Equation (4.149) for elements 1 and N do not have the $\dot{c}_{-1}$ and $\dot{c}_{N+1}$ terms, respectively, and the coefficient of $\dot{c}_n$ is $h/3$. The discretized equations can be placed in the first-order form

$$\mathbf{M}_1 \dot{\mathbf{x}} = \mathbf{M}_2 \mathbf{x} + \mathbf{M}_3 \mathbf{j}, \qquad (4.150)$$

where the state vector $\mathbf{x}(t) = [c_1(t), \ldots, c_N(t)]^T \in R^N$ and the input vector $\mathbf{j}(t) = [j_1(t), \ldots, j_N(t)]^T \in R^N$. The input matrix is diagonal with elements $\mathbf{M}_3(n, n) = a_{2_n} h/2$. The other matrices are block diagonal with elements on the diagonal and off diagonal and the rest of the matrix elements equal to zero. The underlying symmetry of the discretized equations ensures that the eigenvalues will be real. Also, this method has well-defined (monotonic and smooth) convergence properties as the order N increases. The matrix $\mathbf{M}_1$ is positive definite and symmetric and, therefore, nonsingular. The matrix $\mathbf{M}_2$ is also symmetric, but it can be positive *semi*-definite in some cases (e.g., when a_1 is constant) and singular. Boundary conditions other than zero flux can remove this singularity. One can also define the new variable $\Delta c(x, t) = c(x, t) - c(0, t)$, thereby enforcing $\Delta c(0, t) = 0$ and removing the singularity in $\mathbf{M}_2$.

The FEM method allows distributed inputs in the domain and parameters that vary spatially. It is straightforward to add time-varying parameters as well.

For the coupled-domain problem in Section 4.1.1, we define the model parameters as follows:

$$\left.\begin{array}{l} a_{1_n} = D_{\mathrm{m}} \\ a_{2_n} = b \end{array}\right\} x \in (0, L/2) \quad \text{and} \quad \left.\begin{array}{l} a_{1_n} = D_{\mathrm{p}} \\ a_{2_n} = -b \end{array}\right\} x \in (L/2, L), \tag{4.151}$$

where $j_n = I$ over the whole domain. The elements of $\mathbf{M}_1$ must be augmented by ε, as was done for a_1 in Equation (4.147).

4.5.2 Coupled Electrolyte–Solid Diffusion in Li-Ion Electrodes

In Li-ion cells, active material particles are embedded in the porous electrodes. The current density $j(x, t)$, rather than being constant in each electrode, depends on the particle diffusional dynamics. The transfer function between the input current $I(t)$ and the current density distribution takes the form

$$\frac{J(x, s)}{I(s)} = G(x, s), \tag{4.152}$$

taking into account the particle dynamics. In [25], for example,

$$G(x, s) = \frac{v(s)\{\kappa^{\mathrm{eff}} \cosh[v(s)(x/L - 1)] + \sigma^{\mathrm{eff}} \cosh[v(s)(x/L)]\}}{L A \sinh v(s)(\kappa^{\mathrm{eff}} + \sigma^{\mathrm{eff}})}, \tag{4.153}$$

where the dimensionless variable

$$v(s) = L \sqrt{\frac{\kappa + \sigma}{\kappa\sigma\left[\frac{R_{\mathrm{ct}}}{a_{\mathrm{s}}} + \frac{\partial U}{\partial c}\left(\frac{C_{\mathrm{s}}(R, s)}{J(s)}\right)\right]}} \tag{4.154}$$

using the spherical particle transfer function (4.79) or the approximate transfer function (4.108) for $C_{\mathrm{s}}(R, s)/J(s)$.

Taking the Laplace transform of Equation (4.150), we obtain

$$C(s) = (\mathbf{M}_1 s - \mathbf{M}_2)^{-1} \mathbf{M}_3 J(s). \tag{4.155}$$

The input vector for the FEM model (4.155) is related to the distributed transfer function (4.152) as follows:

$$
\mathbf{J}(s) = \begin{bmatrix} J(0,s) \\ J(h,s) \\ \vdots \\ J(L,s) \end{bmatrix} = \begin{bmatrix} G(0,s) \\ G(h,s) \\ \vdots \\ G(L,s) \end{bmatrix} I(s) = \mathbf{G}(s)I(s). \tag{4.156}
$$

Substitution of Equation (4.156) into Equation (4.155) provides the system transfer function:

$$
\frac{\mathbf{C}(s)}{I(s)} = (\mathbf{M}_1 s - \mathbf{M}_2)^{-1}\mathbf{M}_3\mathbf{G}(s). \tag{4.157}
$$

The overall model order is the product of the FEM model order in Equation (4.155) (N) times the order of the current density transfer functions $G((m-1)h, s)$ (all assumed to have the same order equal to N_J) for $m = 1, \ldots, N$. The denominators of the current-density transfer functions are not functions of x, so the poles should not vary with spatial location. Thus, the order can be $N + N_J$ if all of the $G((m-1)h, s)$ have the same denominator.

4.6 Finite-Difference Method

The FDM is the simplest and most commonly used approach to the solution of the diffusion equations found in battery models. As with the FEM, it easily handles spatially varying inputs and parameters. FDM can also be used on nonlinear problems. The method does not always maintain the symmetry of the underlying problem, however, and lacks the convergence guarantees of variation-based (FEM and Ritz) methods. Further information on this method can be found in [17, 18, 26, 27].

In the FDM, the spatial domain $x \in [0, L]$ is discretized into N nodes at $x = 0, h, \ldots,$ $(N-1)h$, where $h = L/(N-1)$ is assumed constant for simplicity. The spatial distribution (e.g., $c(x, t)$) is discretized to nodal values $c_m(t) = c((m-1)h, t)$ for $m = 1, \ldots, N$. Spatial derivatives are approximated by forward-difference

$$
\left.\frac{\partial c}{\partial x}\right|_{x=(m-1)h} \approx \frac{c_{m+1}(t) - c_m(t)}{h}, \tag{4.158}
$$

backward-difference

$$
\left.\frac{\partial c}{\partial x}\right|_{x=(m-1)h} \approx \frac{c_m(t) - c_{m-1}(t)}{h}, \tag{4.159}
$$

or central-difference

$$
\left.\frac{\partial c}{\partial x}\right|_{x=(m-1)h} \approx \frac{c_{m+1}(t) - c_{m-1}(t)}{2h} \tag{4.160}
$$

formulas. Forward difference and backward difference are used at the left and right boundaries, respectively, because c_0 and c_{N+1} do not exist. Central difference is used for the remaining nodes, typically for second derivatives:

$$\left.\frac{\partial^2 c}{\partial x^2}\right|_{x=(m-1)h} \approx \frac{c_{m+1}(t) + c_{m-1}(t) - 2c_m(t)}{h^2}. \tag{4.161}$$

4.6.1 Electrolyte Diffusion

The FDM discretized field equation for electrolyte diffusion is

$$\dot{c}_m = a_1 \frac{c_{m+1} + c_{m-1} - 2c_m}{h^2} + a_2 j_m. \tag{4.162}$$

The zero flux boundary conditions

$$\left.\frac{\partial c}{\partial x}\right|_{x=0} = \frac{c_2(t) - c_1(t)}{h} = 0, \tag{4.163}$$

$$\left.\frac{\partial c}{\partial x}\right|_{x=L} = \frac{c_N(t) - c_{N-1}(t)}{h} = 0 \tag{4.164}$$

mean that $c_1(t) = c_2(t)$ and $c_N(t) = c_{N-1}(t)$, so the model order is $N - 2$. The discretized equations (4.162) can be written state-variable form with $\mathbf{x}(t) = [c_2, \ldots, c_{N-1}]^T \in R^{N-2}$ and input vector $\mathbf{u}(t) = [j_2(t), \ldots, j_{N-1}(t)]^T$. The state matrices

$$\mathbf{A} = \frac{1}{h^2}
\begin{bmatrix}
-a_1 & a_1 & 0 & \cdots & 0 & 0 & 0 \\
a_1 & -2a_1 & a_1 & \cdots & 0 & 0 & 0 \\
0 & a_1 & -2a_1 & \cdots & 0 & 0 & 0 \\
 & & & \ddots & & & \\
 & & & & -2a_1 & a_1 & 0 \\
 & & & & a_1 & -2a_1 & a_1 \\
0 & 0 & 0 & \cdots & 0 & a_1 & -a_1
\end{bmatrix}, \tag{4.165}$$

with $\mathbf{B} = \mathrm{diag}(\{a_2, a_2, \ldots, a_2\})$, $\mathbf{C} = [0, \ldots, 0, 1]$ and $\mathbf{D} = 0$. In general, the coefficients a_1 and a_2 can be spatially dependent. For the coupled-domain problem in Section 4.1.1, we define the model parameters as in the FEM method in (4.151) and divide each row of $\mathbf{A}$ and $\mathbf{B}$ by ε. The $\mathbf{A}$ matrix is symmetric for the single-domain problem but not the coupled problem. In the coupled-domain problem, asymmetry appears in the $\mathbf{A}$ matrix at the interface between the two domains.

4.6.2 Nonlinear Coupled Electrolyte–Solid Diffusion in Pb Electrodes

The coupled electrolyte–solid diffusion model for Pb–acid cells presented in Section 4.1.2 includes several linearizing assumptions. In this section we retain a nonlinear term in the governing equations and develop the discretized, nonlinear ODEs using the FDM. It would be quite difficult to generate nonlinear ODEs for this example using any of the other previously developed methods.

The model includes the field equations

$$\dot{c} = a_1 c'' + a_2 \phi'',$$ (4.166)

$$\phi'' = a_3 \phi - a_4^* \ln(c)'',$$ (4.167)

where the term $a_4^* \ln(c)''$ has replaced the linearized term $a_4 c''$ in the potential field equation (4.167). The boundary conditions, output voltage equation, and solid-phase potential equation remain the same.

The domain is discretized into N nodes with constant length h. The distributed concentration and potential are approximated by their nodal values $c_m(t) = c((m-1)h, t)$ and $\phi_m(t) = \phi((m-1)h, t)$ for $m = 1, \ldots, N$. The concentration field equation (4.166) is discretized using central difference to produce

$$\dot{c}_m = a_1 \frac{c_{m+1} + c_{m-1} - 2c_m}{h^2} + a_2 \frac{\phi_{m+1} + \phi_{m-1} - 2\phi_m}{h^2}.$$ (4.168)

Field equation (4.167) discretizes to

$$\frac{\phi_{m+1} + \phi_{m-1} - 2\phi_m}{h^2} = a_3 \phi_m - a_4^* \frac{\ln(c_{m+1}) + \ln(c_{m-1}) - 2\ln(c_m)}{h^2}$$ (4.169)

or

$$\frac{1}{h^2}(\phi_{m+1} + \phi_{m-1}) - \left(\frac{2}{h^2} + a_3\right)\phi_m = \frac{a_4^*}{h^2} \ln\left(\frac{c_{m+1}c_{m-1}}{c_m^2}\right).$$ (4.170)

The zero-flux concentration boundary conditions require $c_1(t) = c_2(t)$ and $c_{N-1}(t) = c_N(t) = 0$ so the state vector $\mathbf{x}(t) = [c_2, \ldots, c_{N-1}]^T$. The potential boundary conditions require

$$\phi'(0, t) = \frac{\phi_2 - \phi_1}{h} = a_5 I,$$ (4.171)

$$\phi'(L, t) = \frac{\phi_{N-1} - \phi_N}{h} = a_5 I.$$ (4.172)

We define the potential vector as $\boldsymbol{\phi} = [\phi_2, \ldots, \phi_{N-1}]^T$. Equations (4.168) and (4.170) can be rewritten in matrix form as

$$\dot{\mathbf{x}} = \mathbf{A}\mathbf{x} + a_2 \mathbf{B}\boldsymbol{\phi},$$ (4.173)

$$\mathbf{P}\boldsymbol{\phi} = \mathbf{f}(\mathbf{x}),$$ (4.174)

where $\mathbf{A}$ is given in Equation (4.165) and

$$\mathbf{B} = \frac{1}{h^2} \begin{bmatrix} -1 & 1 & 0 & \cdots & 0 & 0 & 0 \\ 1 & -2 & 1 & \cdots & 0 & 0 & 0 \\ & & & \ddots & & & \\ & & & & 1 & -2 & 1 \\ 0 & 0 & 0 & \cdots & 0 & 1 & -1 \end{bmatrix}, \tag{4.175}$$

and $\mathbf{P} = \mathbf{B} - a_3 \mathbf{I}$, where $\mathbf{I}$ is the $N - 2 \times N - 2$ identity matrix. The nonlinear vector function

$$\mathbf{f}(\mathbf{x}) = \frac{a_4^*}{h^2} \begin{bmatrix} \ln\left(\frac{c_3}{c_2}\right) \\ \ln\left(\frac{c_4 c_2}{c_3^2}\right) \\ \vdots \\ \ln\left(\frac{c_{N-2}}{c_{N-1}}\right) \end{bmatrix}. \tag{4.176}$$

The output voltage

$$V(t) = \phi_s(L, t) - \phi_s(0, t) = a_7 \int_0^x \int_0^{x_2} \phi \, dx_1 \, dx_2 \tag{4.177}$$

is formed by Euler integration of the $\phi(x, t)$ distribution.

4.7 System Identification in the Frequency Domain

It is often relatively easy to determine the transcendental transfer function and/or frequency response of a battery system. In simple cases, the transcendental transfer function can be derived analytically. The two-domain diffusion and cylindrical particle transfer functions, for example, are derived earlier in this chapter and shown in Equations (4.63) and (4.78), respectively. Transcendental transfer functions can include non-polynomial functions such as hyperbolics and square roots, so it is difficult to produce a standard transfer function in the form of a ratio of two polynomials in s.

In this section we explore the use of system identification techniques to extract a linear, discretized model based on frequency-response data. System identification has a long history and background information can be found in many texts, including [28] and [29]. In this case, we are trying to find the lowest order transfer function that best matches a fixed set of frequency-response data. The best match is in a least-squares sense – it minimizes the sum of the squares of the errors between the data and the model. Fortunately, there are many least-square optimizers that can be applied to this problem. This section provides a short introduction to least-squares optimization. The reader may refer to one of the many excellent books on optimization (e.g., [30], [31], and [32]). We constrain our approach to the diffusion models that underly battery dynamics. The eigenvalues are real and we assume they are not repeated. This simplifies the system identification problem and allows relatively simple optimizers to be used. Also, as the frequency-response data are generated by mathematical models, they are free from measurement and input noise, so stochastic analysis is not required. Many of the techniques developed in this section, however, can be applied to measured frequency-domain or time-domain data to extract a model that best matches the experimental data.

4.7.1   System Model

For diffusion systems with real poles and residues, we seek a linear model of the form

$$\hat{G}(\theta, s) = \sum_{k=1}^{N} \frac{R_k}{s - p_k}, \tag{4.178}$$

where the model order N is given and the residues R_k and poles p_k are unknown. The parameter vector $\theta = [R_1, \ldots, R_N, p_1, \ldots, p_N]$. For stability, the poles are all negative. The residues, however, can take on either sign. Equation (4.178) only applies to single-input single-output (SISO) transfer functions, but models with multiple inputs and/or outputs that have transfer functions with the same poles will also be addressed.

4.7.2   Least-Squares Optimization Problem

The cost function to be minimized by the optimization algorithm is the sum of the squares of the errors between the complex frequency-response data $G(i\omega_j)$ and the estimate $\hat{G}(\theta, i\omega_j)$,

$$e(\theta, \omega_j) = G(i\omega_j) - \hat{G}(\theta, i\omega_j), \tag{4.179}$$

where the frequency-response data is provided at $j = 1, \ldots, N_{\text{eval}}$ frequencies with $N_{\text{eval}} > N/2$. The cost function

$$\text{CF} = \sum_{j=1}^{N_{\text{eval}}} ([\Re\{e(\theta, \omega_j)\}]^2 + [\Im\{e(\theta, \omega_j)\}]^2), \tag{4.180}$$

where $\Re$ and $\Im$ indicate the real and imaginary parts, respectively. The objective is to find θ that minimizes CF.

The residuals are the error terms that are squared in Equation (4.180):

$$r_m(\theta) = \Re\{e(\theta, \omega_j)\} = \Re\{G(i\omega_j)\} + \sum_{k=1}^{N} \frac{R_k p_k}{p_k^2 + \omega_j^2}$$

$$\text{for } m = 1, \ldots, N_{\text{eval}} \tag{4.181}$$

$$r_m(\theta) = \Im\{e(\theta, \omega_j)\} = \Im\{G(i\omega_j)\} + \sum_{k=1}^{N} \frac{R_k \omega_j}{p_k^2 + \omega_j^2}$$

$$\text{for } m = N_{\text{eval}} + 1, \ldots, 2N_{\text{eval}}$$

where the first N_{eval} are associated with the real parts of the error and the last N_{eval} with the imaginary parts. Equation (4.181) shows that the residuals are linear in R_k but nonlinear in p_k.

The Jacobian, $\mathbf{J} = [\mathbf{J}_R, \mathbf{J}_P] \in \mathfrak{R}^{2N_{\text{eval}} \times 2N}$, is the gradient of the residuals of the cost function with respect to the model parameters θ. Considering R_j as the only unknown parameter, the Jacobian

$$\mathbf{J}_R(k, j) = \frac{\partial r_k}{\partial R_j}$$

$$= \begin{cases} \frac{p_j}{p_j^2 + \omega_k^2} & \text{for } k = 1, \dots, N_{\text{eval}} \\ \frac{\omega_k}{p_j^2 + \omega_k^2} & \text{for } k = N_{\text{eval}} + 1, \dots, 2N_{\text{eval}} \end{cases} \quad \text{for } j = 1, \dots N \tag{4.182}$$

is in $\mathfrak{R}^{2N_{\text{eval}} \times N}$ and independent of R_j but dependent on p_j. For p_j, the Jacobian

$$\mathbf{J}_P(k, j) = \frac{\partial r_k}{\partial p_j}$$

$$= \begin{cases} \frac{R_j \left(\omega_k^2 - p_j^2 \right)}{\left(p_j^2 + \omega_k^2 \right)^2} & \text{for } k = 1, \dots, N_{\text{eval}} \\ -\frac{2R_j p_j \omega_k}{\left(p_j^2 + \omega_k^2 \right)^2} & \text{for } k = N_{\text{eval}} + 1, \dots, 2N_{\text{eval}} \end{cases} \quad \text{for } j = 1, \dots N \tag{4.183}$$

is also in $\mathfrak{R}^{2N_{\text{eval}} \times N}$.

Solving for R_k if p_k is known constitutes a linear least-squares problem because the Jacobian J_R is independent of R_k. We can rewrite the residuals as

$$\mathbf{r} = \begin{bmatrix} r_1 \\ \vdots \\ r_{2N_{\text{eval}}} \end{bmatrix} = \mathbf{g} - \mathbf{J}_P \mathbf{R}, \tag{4.184}$$

where $\mathbf{R} = [R_1, \dots, R_N]'$, $\mathbf{g}(k) = \mathfrak{R}\{G(\omega_k)\}$ for $k = 1, \dots, N_{\text{eval}}$ and $\mathbf{g}_k = \mathfrak{I}\{G(\omega_k)\}$ for $k = N_{\text{eval}} + 1, \dots, 2N_{\text{eval}}$. The cost function then becomes

$$\text{CF} = |\mathbf{g} - \mathbf{J}_P \mathbf{R}|^2. \tag{4.185}$$

The cost function in Equation (4.185) is convex, so the global minimum is at

$$\frac{\partial \text{CF}}{\partial \mathbf{R}} = 2\mathbf{J}^{\text{T}}(\mathbf{g} - \mathbf{JR}) = 0, \tag{4.186}$$

producing the normal equations

$$\mathbf{JJ}^{\text{T}} \mathbf{R}^* = \mathbf{J}^{\text{T}} \mathbf{g}, \tag{4.187}$$

where $\mathbf{R}^*$ is the global minimizer of CF. If $\mathbf{JJ}^{\text{T}}$ is invertible then we can solve Equation (4.187) directly for $\mathbf{R}^*$.

For the nonlinear least-squares problem associated with finding the poles **p** there are many algorithms that can be used. Gauss–Newton uses a line search and an effective approximation to the Hessian. The Levenberg–Marquardt method uses the same Hessian approximation but uses a trust region strategy rather than line search. The analytical Jacobian is available, so it makes sense to choose an optimizer that can use the gradient information.

4.7.3   Optimization Approach

Matlab provides a variety of tools that can be applied to system identification using a frequency-domain approach. The basic Matlab package has some optimizers. The Optimization Toolbox has more sophisticated optimizers that are more applicable to the System ID problem.

Matlab includes `fminsearch.m` and `fminbnd.m` in its basic package. These allow multidimensional unconstrained nonlinear minimization using a Nelder–Mead direct search method (`fminsearch.m`) and single-variable bounded nonlinear function minimization (`fminbnd.m`). In general, we are searching for more than one pole, so `fminbnd.m` is not useful. Nelder–Mead is a derivative-free optimization method that does not take advantage of the analytical Jacobian in this problem.

The Optimization Toolbox in Matlab provides a variety of other optimizers, including `lsqnonlin.m` – an algorithm that is explicitly designed for nonlinear least-squares problems and can use the analytical Jacobian. Lower and upper bounds can be included on the design variables, so we can constrain the poles to be negative. The default optimizer is considered to be "Large Scale" and requires that $2N_{eval} > N$. The "Large Scale" algorithm is a subspace trust-region method based on the interior-reflective Newton method. A "Medium Scale" option can also be selected that uses a Levenberg–Marquardt method but cannot provide bounds on the design variables. The `lsqnonlin.m` algorithm can be tuned by choosing input parameters such as MaxFunEvals, MaxIter, TolFun, and TolX. These parameters are provided with default values at startup but can be specified by the user as well.

Two input parameters to `lsqnonlin.m` are most critical: TolFun and TolX. In most cases, the optimizer terminates when the

- Change in x is less than the specified tolerance (TolX).
- Change in the residual is less than the specified tolerance (TolFun).
- Number of iterations is exceeded (MaxFunEvals or MaxIter).

The TolX must be 10 or 100 times smaller than the slowest pole (or lowest frequency) that is expected in the system response. TolFun can be based on the smallest magnitude that the optimized model should capture.

The `lsqnonlin.m` algorithm will find a local minimum, but there is no guarantee that this is the global minimum or best-fit solution. The nonlinear optimization problem associated with finding the poles is not convex and there can be a large number of closely spaced minima. The poles are interchangeable; so, if two poles switch the error will be unchanged. The minimum that `lsqnonlin.m` returns is very sensitive to the initial guess that is required as input to the algorithm. One can wrap an outer loop around the optimizer that manipulates the initial guesses using genetic, stochastic, or global minimization schemes. Alternatively, these problems typically run quite quickly on even a desktop PC for the low-order approximations ($N < 10$) that are sought for systems analysis. Thus, an outer loop that simply saturates the initial guess space is a simple and effective means of finding a global minimum. It is always advisable to check the optimization scheme using a known transfer function to see whether the solution converges to the known global minimum.

The overall optimization approach has the following steps:

1. Pick $\omega_{\min}$, $\omega_{\max}$, N_{eval}, and N. The frequencies $\omega_{\min}$ and $\omega_{\max}$ set out the frequency bandwidth of interest. Subtract out any integrators in the frequency response so that the DC response is constant. Set TolX and TolFun.
2. Pick initial guesses for poles such that $p_1 > p_2 > \cdots > p_N$ with all $p_k < 0$. A Monte Carlo approach may be used or a more systematic marching through the design space to pick initial guesses. A log-scale distribution for initial guesses ensures that they cover the desired bandwidth. Enforcing a minimum separation between poles can minimize singularity problems in the optimizer.
3. Use lsqnonlin.m to minimize the cost function (4.186). For each iteration, solve the linear least-squares problem (4.187) for $R_1, \ldots, R_N$.
4. Return to step 2 until a sufficient number of initial guesses N_{guess} have been used to saturate the design space.
5. Solution with minimal error out of N_{guess} initial guesses is the "global" minimum.

4.7.4 Multiple Outputs

Expanding the approach described in the previous section to multiple output systems has many applications in battery system engineering, even the most sophisticated battery models with one input (current) and one output (voltage). Regardless of how the frequency response is generated, the method produces an optimal linear model. Systems engineers are often interested in the conditions inside the cell, requiring a model with multiple outputs. The optimization problem can be set up to include multiple outputs, but the order can become quite high and finding the global minimum difficult. Alternatively, one can base the multiple output model on the single output optimization results. For battery cell models, the global optimum model for the current-to-voltage transfer function (impedance) can be found. The internal cell dynamics presumably share the same poles. Internal dynamics that do not share the same poles will most likely be unobservable and/or uncontrollable. If we assume that the internal dynamics share the same poles as impedance, then all of the outputs have the same transfer-function denominator. To capture the internal state dynamics requires only the simple linear least-squares problem to determine the residues (and hence numerator) for a given output. From a state-space perspective, the multiple output system will have the same $\mathbf{A} = \text{diag}(p_1, \ldots, p_N)$ and $\mathbf{B} = [1, \ldots, 1]^{\mathsf{T}}$ matrices. The output matrix $\mathbf{P}$ will have rows corresponding to the residues from each of the multiple outputs. This single-input multiple-output model will be low order and automatically controllable and observable, providing a firm foundation for battery estimation and control.

4.7.5 System Identification Toolbox

The System Identification Toolbox is powerful, sophisticated, and specifically designed to handle the identification of general transfer functions from frequency-response data. The pem.m function, for example, computes the prediction error estimate of a general linear model. The input data can be in the form of an identified frequency response data (IDFRD) model and the output is the identified linear model. Using the IDGREY model object, the user can define a linear model structure in the pole/residue form of Equation (4.178) or search for a more general linear model structure in state space or polynomial form. Providing pem.m with multiple output data produces a multiple-output linear model.

4.7.6   Experimental Data

The frequency response fitting algorithms used to fit transcendental transfer functions can also be used with experimental data to empirically fit a model. The model would only fit the measured impedance data and not be able to predict the internal states of the battery, however. This approach can produce accurate models for a specific battery and a specific stage in its life, but it cannot be used to predict SOC or other internal states of the battery. The relationships between the model and the underlying physical processes and governing equations are difficult to ascertain using an empirical approach.

Problems

4.1   To decouple the three domains (negative electrode, separator, and positive electrode) in an Li-ion cell, we derived an approximate boundary condition at the electrode–separator interface in Problem 3.9 as

$$\frac{\partial c_e}{\partial x}\bigg|_L = -\alpha c_e(L, t), \tag{A.4.1}$$

where $\alpha = 2/\delta_{sep}$. The other boundary condition is $\frac{\partial c_e}{\partial x}(0, t) = 0$. The field equation is

$$\frac{\partial c_e}{\partial t} = a_1 \frac{\partial^2 c_e}{\partial x^2} + a_2 j_{avg}(t), \tag{A.4.2}$$

assuming a constant current density distribution. In this problem, we study the positive electrode $x = 0 \cdots L$ which is decoupled from the other two domains.
   (a)   Determine the characteristic equation for this system, the roots of which are the eigenvalues λ. Determine a simplified equation for the fast eigenvalues as $\lambda \to -\infty$.
   (b)   Derive the transcendental transfer function $\frac{C_e(x,s)}{J_{avg}(s)}$ using the analytical method.
   (c)   Derive a discretized transfer function using IMA with a parabolic concentration distribution.
   (d)   Derive a state-space model using the Ritz method and polynomial admissible functions.
4.2   Equation (4.108) is the first-order transfer function for an active material particle discretized using IMA.
   (a)   Rederive the transfer function using a volume $\int dV$ average in the IMA discretization rather than the radial $\int dr$ average used in that analysis.
   (b)   Derive the first-order, IMA discretized, volume-averaged transfer function for a cylindrical particle.

4.3   For the single-domain diffusion problem

$$\frac{\partial c}{\partial t} = a_1 \frac{\partial^2 c}{\partial x^2} + a_2 j$$

with the boundary conditions

$$\left.\frac{\partial c}{\partial x}\right|_{x=0} = \left.\frac{\partial c}{\partial x}\right|_{x=L} = 0,$$

the current distribution is often concentrated near the separator as follows: $j(x, t) = \alpha x^2 I(t)$. For this model developed, the input is $I(t)$ and the output is $c(L, t)$.
(a)   Derive the analytical transfer function $C(L, s)/I(s)$.
(b)   Derive the analytical state-space matrices using an eigenvalue approach.
(c)   Derive the state-space matrices for an FDM discretization of this problem.
(d)   Derive the state-space matrices for an FEM discretization of this problem.
4.4   For solid-state diffusion in cylindrical particles, the transfer function derived in this chapter (normalized for simplicity) is

$$G(s) = \frac{I_0(\Gamma(s))}{\Gamma(s) I_1(\Gamma(s))},$$

where $\Gamma(s) = R\sqrt{-s/D}$, $I_0(s)$ and $I_1(s)$ are zeroth- and first-order modified Bessel functions of the first kind. Find the second-order Padé approximation of $sG(s)$.
4.5   Solid-state diffusion in spherical particles is governed by

$$\frac{\partial c}{\partial t} = D\left(\frac{\partial^2 c}{\partial r^2} + \frac{2}{r}\frac{\partial c}{\partial r}\right) \quad \text{for } r \in (0, R)$$

with boundary conditions

$$\left.\frac{\partial c}{\partial r}\right|_{r=0} = 0 \text{ and } \left.\frac{\partial c}{\partial r}\right|_{r=R} = \frac{-j}{DAF}.$$

Discretize this problem using the Ritz method and polynomial admissible functions $C_m(r) = x^m$.
4.6   Discretize the transfer function

$$G(s) = -\frac{2a_2\alpha L}{s\beta}\frac{\cosh \beta L}{\sinh \beta L} + \frac{a_2\alpha(2a_1 + L^2 s)}{s^2} \tag{A.4.3}$$

using a third-order Padé approximation, where $\beta = \sqrt{s}/\sqrt{a_1}$. Choose the numerator order of the Padé approximation to match (if possible) the high-frequency asymptote of the transfer function.

5

System Response

The state-space and transfer-function models developed in Chapter 4 are used to predict the output response of battery systems to different inputs. The outputs can include concentration and potential distributions or terminal voltage. The input to the battery model is current and the output is voltage. In some cases, the power from the pack is prescribed input. This can only be solved in a feedback fashion where the current input is adjusted until power (i.e., current times voltage) is at the desired level. A simpler method is to assume a constant terminal voltage and convert the power profile directly to a current profile.

Battery systems engineers are interested in the response of the battery pack to step (constant current) charge and discharge, sinusoidal, and duty-cycle currents. The step current response is often expressed at a C rate. A $1C$ charge means that the battery will be completely charged from zero capacity in 1 h. Constant charge and discharge currents are used to characterize the battery voltage response under different C rates.

The frequency response resulting from sinusoidal currents is most important over the bandwidth of interest for the battery system, typically around 10 Hz. The frequency response can be generated experimentally using electrochemical impedance spectroscopy (EIS), where small sinusoidal currents are injected into the battery and the output voltage magnitude and phase are calculated. The frequency response to sinusoidal current inputs provides wide bandwidth information about the dynamic response of the battery.

Duty cycles are defined for the specific application, typically consisting of alternating charge and discharge pulses. A variety of duty cycles have been developed for hybrid vehicles, many of which are adopted from the Environmental Protection Agency (EPA) vehicle speed versus time duty cycles that are used to assess fuel efficiency. Examples include the Federal Urban Driving Schedule (FUDS), SFUDS, and DST set by the US Advanced Battery Consortium [33] and the Hybrid Pulse Power Characterization Test (HPPC) set by the Partnership for a New Generation of Vehicles [34]. These HEV testing cycles are used to test batteries for performance and life.

The models developed in Chapter 4 use analytical, integral approximation, Ritz, Padé, finite element, finite difference, and system identification methods to discretize the underlying PDEs. For systems design and analysis it is essential that the model has a fast convergence rate so that it can accurately predict the response with a minimum number of states. The estimators

Battery Systems Engineering, First Edition. Christopher D. Rahn and Chao-Yang Wang.
© 2013 John Wiley & Sons, Ltd. Published 2013 by John Wiley & Sons, Ltd.

and battery management algorithms presented in the next chapters are based on these models and should be accurate but able to run in real time. Real-time operation typically requires the entire battery model to be less than 10 states. In this section, we compare the methods for "prediction efficiency" (accuracy versus model order) using the step and frequency response.

There are two ways to obtain the low-order ($N < 10$) models that can be practically used for real-time, model-based estimation and control. First, one can use an approximation method that is efficient at accurately capturing the system response with a few states. In this chapter we will characterize the prediction accuracy of the various discretization methods as a function of the model order. Second, one can generate an accurate high-order model and use model-order reduction techniques to reduce the number of states in the model but retain response prediction accuracy. Model-order reduction techniques are introduced in this chapter to achieve this goal. To compare the various discretization and model-order reduction techniques for numerical efficiency, model error metrics are introduced for both the step and frequency responses based on L_2 and L_∞ norms.

Reduced-order models of battery cells can be represented by equivalent circuits [35]. If the underlying equations are linear, then equivalent circuit or equivalent mechanical models can often be derived. These equivalent models can provide insight into the battery behavior and electrical/mechanical interpretation of the frequency response and time response of electro-chemical batteries. They can also allow engineers to integrate them with existing software that simulates electronic circuits or mechanical systems. Our goal, however, is to integrate the battery models into a systems engineering environment such as Matlab to allow model-based estimation and control. Transfer-function and state-space models are the tools of the systems engineering trade.

5.1 Time Response

Battery systems respond in real time to current charge and discharge inputs. The time response can be characterized by the step response. Figure 5.1 shows example experimental data for a fully charged LiFePO$_4$ battery undergoing pulse testing. The current cycles between 30 s of discharge, 30 s of rest, 30 s of charge, and 30 s of rest. For the first minute, current pulses of $2C$ are applied, then $5C$, and finally $10C$. At the end of each cycle, the battery voltage returns to near the initial value because the same amount of current goes in and out of the battery during the cycle. The SOC at the beginning and end of the cycle is the same. More energy goes into charging the battery, however, than comes out during discharge. Although the current is the same during charge and discharge, the charging voltage is higher than the discharging voltage. This means that the charging power is higher than the discharging power, as one would expect.

The experimental results demonstrate that this battery responds linearly over this range of applied currents. The positive and negative current pulses generate the same response except reversed in sign. The different amplitude current pulses produce responses that are scaled by the input amplitude with the same shape. These are characteristics of linear systems. Even at the high current rate, the pulse duration is not long enough to deplete the battery charge such that nonlinearities become important. If the SOC drops to much lower levels or the battery is overcharged, however, nonlinearities become significant. Fortunately, HEVs are designed to maintain the SOC in a desired operating range, so linear models can often be used. PHEVs with their larger SOC operating range, however, may require nonlinear models to accurately predict the battery response.

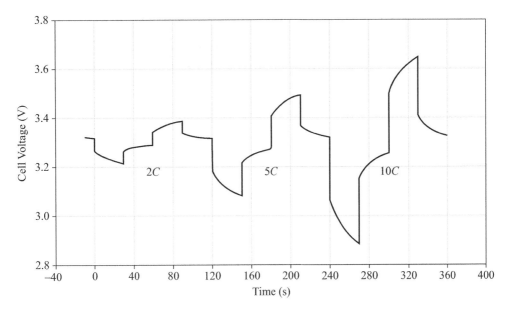

Figure 5.1 Example pulse test experimental data for LiFePO$_4$ battery showing repeated discharge–charge cycles at 2C, 5C and 10C

Evaluation of the accuracy with which a model predicts the battery response requires the definition of an error metric. In this chapter we consider the effects of discretization methods and model-order reduction techniques on the accuracy of the discretized and reduced-order model relative to the exact solution.

5.1.1 Constant Charge/Discharge Current

The step response to a constant charge or discharge current is often used to characterize the performance of batteries. The capacity can be determined by the area under the voltage versus time curve. The voltage response and the effects of high rate of charge and discharge can be investigated. For short times and relatively low current rates, linear models can accurately predict the response. High current rates and long discharge/charge times can lead to nonlinear effects as the SOC reaches low/high values during undercharge/overcharge operation. In this section we study the step response of several battery models developed in Chapter 4. The steady-state response as $t \to \infty$ and transient response are calculated. The effects of discretization on the model accuracy are quantified.

Steady-State Response

Battery models with voltage output and current input typically have a pole at the origin. This means that an impulsive current input results in a steady-state change in voltage but a step current input causes a linearly increasing (charge) or decreasing (discharge) voltage response.

If the current is applied for a long enough time and/or at a high enough rate, then eventually the voltage response flattens out due to nonlinearities or unmodeled side reactions.

Consider, for example, the single-domain diffusion problem discussed in Section 4.1.1. The transfer function from boundary current flux to concentration is

$$\frac{C(L, s)}{I(s)} = \frac{a_3 \cosh(\beta L)}{\beta \sinh(\beta L)} = G(s), \qquad (5.1)$$

where $\beta = \sqrt{s/a_1}$. For a unit impulse current input, $I(s) = 1$, and using the final value theorem

$$y(\infty) = \lim_{s \to 0} sY(s) \qquad (5.2)$$

we obtain

$$c(L, \infty) = \lim_{s \to 0} sG(s)I(s) = \lim_{s \to 0} \frac{a_3 s \cosh\left(\sqrt{L^2 s/a_1}\right)}{\sqrt{s/a_1} \sinh\left(\sqrt{L^2 s/a_1}\right)} = \frac{a_1 a_3}{L}. \qquad (5.3)$$

Thus, the steady-state change in concentration is proportional to the rate of diffusion a_1 and current input multiplier a_3 and inversely proportional to the length of the domain L. For a step current input, the concentration continues to build as current fluxes into the domain and there is no steady state.

The initial value theorem,

$$y(0^+) = \lim_{s \to \infty} sY(s), \qquad (5.4)$$

can be used to calculate the initial value resulting from different inputs. For the single-domain diffusion problem, the initial response to an impulse can be calculated from Equation (5.3) and is found to be infinite. The input feeds through directly to the output, causing an infinitely high (and infinitely thin) impulse in the output. For a unit step input, $I(s) = 1/s$, however,

$$c(L, 0^+) = \lim_{s \to \infty} sG(s)I(s) = \lim_{s \to \infty} \frac{a_3 \cosh\left(\sqrt{L^2 s/a_1}\right)}{\sqrt{s/a_1} \sinh\left(\sqrt{L^2 s/a_1}\right)} = 0, \qquad (5.5)$$

so the step response starts at zero concentration and, as $t \to \infty$, the concentration decreases linearly with time under discharge, never reaching steady state.

It is interesting to note that the IMA for this problem matches the impulse response final value in Equation (5.4), predicts an infinite initial value for an impulse input, but incorrectly estimates a nonzero initial value for a step input. Thus, the IMA introduces inaccuracies in predicting the steady-state response of this simple diffusion problem.

The coupled-domains problem of electrolyte diffusion introduced in Section 4.1.1 has different steady-state characteristics than the single-domain problem. Using the final value theorem, the steady-state concentration response to a unit step current input is

$$
c(L, \infty) - c(0, \infty) = \frac{b\varepsilon_p}{D_p} \lim_{s \to 0} \left\{ \frac{\begin{array}{c} 4\alpha \sinh(\frac{1}{2}\beta_p L) - 2(\zeta - \alpha)\sinh(\beta_p \gamma_1) \\ + 4\zeta \sinh(\frac{1}{2}\alpha\beta_p L) - 2(\zeta + \alpha)\sinh(\beta_p \gamma_2) \end{array}}{\beta_p^2 [(\zeta - \alpha)\sinh(\beta_p \gamma_1) + (\zeta + \alpha)\sinh(\beta_p \gamma_2)]} \right\}
$$

$$
= -\frac{b\varepsilon_m \varepsilon_p L^2 (D_m + D_p)}{4 D_m D_p (\varepsilon_m + \varepsilon_p)}, \tag{5.6}
$$

using the definitions of γ_1 and γ_2 from Equation (4.62) and $\beta_p = \sqrt{\varepsilon_p s / D_p}$.

Equation (5.6) shows that the concentration change from a constant positive discharge current is negative and proportional to the length of the domain squared and inversely proportional to the effective diffusion coefficient. The L^2 in Equation (5.6) results from the distributed input for this model, unlike the L in Equation (5.4) that results from the boundary input in the single-domain problem. Also, in the single-domain problem, the output was the concentration at $x = L$, rather than the difference between the concentration at $x = L$ and $x = 0$ used in this problem. The effective diffusion coefficient results from the two diffusional conductances in series. Faster diffusion (higher D_m/ε_m and D_p/ε_p) causes less gradient in the concentration, so $c(L, \infty)$ is closer to $c(0, \infty)$ and the output is smaller. If either diffusional conductance approaches zero then the concentration builds rapidly in the associated electrode and the output increases. With larger L, the ions take longer to traverse the domain, resulting in larger concentration gradients. Using the initial value theorem, the initial output to a step input is $c(L, 0^+) - c(0, 0^+) = \lim_{s \to \infty} G(s) = 0$, so the step response starts at zero and converges to a finite steady-state value.

The analytical solution also allows the calculation of the steady-state concentration distribution for the distributed output $\Delta c(x, t) = c(x, t) - c(0, t)$. Application of the final value theorem to the exact solution in the two domains results in

$$
\frac{4 D_p (\zeta - 1)}{b\varepsilon_p L^2} \Delta c(x, \infty) = \begin{cases} -4\alpha^2 x^{*2} & \text{for } x^* \in (0, 1/2) \\ 3\zeta - \alpha^2 + 4\zeta x^{*2} - 8\zeta x^* & \text{for } x^* \in (1/2, 1) \end{cases}, \tag{5.7}
$$

where $x^* = x/L$. The steady-state solutions from Equation (5.7) are plotted in Figure 5.2. The zero-flux boundary conditions at $x = 0$ and $x = L$ are satisfied. The slope at $x = L/2$ changes if the two domains have different diffusion constants ($\alpha \neq 1$) or electrode phase volume fractions ($\zeta \neq 1$). The concentration is positive for $x \in (0, L/2)$ and negative in $x \in (L/2, L)$ owing to the opposite signs of transfer current density in the two domains.

Transient Response

The transient response of a battery cell to step changes in charge/discharge current reveals how the concentration, potential, current density, and terminal voltage change with time under constant-current loading. In this section, the step response of the electrolyte diffusion model is calculated using parameters that are typical of battery cells. Overall cell level outputs and distributions through the cell are calculated as a function of time. The case for discharge step

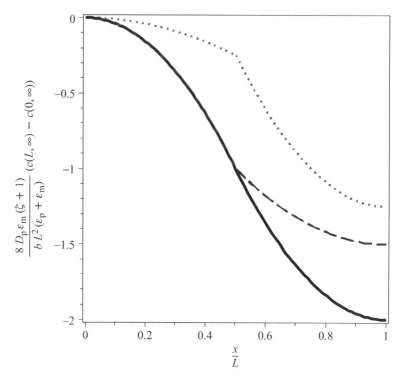

Figure 5.2 Steady-state concentration distributions for coupled electrolyte diffusion model: $\alpha = 0.5$ and $\zeta = 1$ (dashed), $\alpha = \zeta = 1$ (solid), and $\alpha = 1$ and $\zeta = 0.5$ (dotted)

input currents is shown here, but, because the system is linear, the charge response simply has the opposite sign.

Table 5.1 shows the parameters for the electrolyte diffusion model. Batteries typically pack the cells in very small volumes. The length of the domain of less than a tenth of a millimeter is typical of Li-ion cells. The eigenvalues or poles of the analytical transfer function start at 0.14 rad/s, corresponding to a time constant of 7.1 s. There are 26 eigenvalues below 100 rad/s (16 Hz). The residues start at -1.05 and decrease with increasing frequency. The 26th residue is almost zero. The odd residues (1, 3, etc.) are generally several orders of magnitude smaller than the even residues.

Table 5.1 Electrolyte diffusion model parameters

Parameter	Value
L (μm)	100
t_0	0.363
A (cm^2)	10 452
D^{ref} (cm^2/s)	2.6×10^{-6}
ε_m	0.332
ε_p	0.28

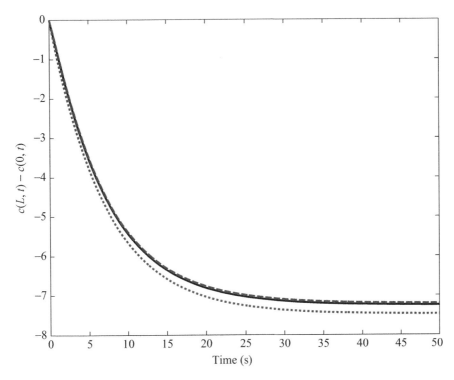

Figure 5.3 Discharge step response for electrolyte diffusion. Analytical solution with 26 (solid black), 4 (dashed), and 2 (dotted) term approximations; output concentration ($c(L, t) - c(0, t)$) in mol/m^3

The discharge step response is shown in Figure 5.3 for different truncation orders. The output is the difference in concentration across the cell $c(L, t) - c(0, t)$. The initial concentration is zero and current fluxes into the anode and fluxes out of the cathode, creating a negative change in relative concentration. The time response settles out in approximately five time constants at 35 s to a steady-state value as predicted by the steady-state analysis in the previous section. As the model order increases from 2 to 4 to 26 modes, the response converges. Even with only two modes the response is fairly accurate. A truncated analytical solution is an efficient discretization method for the electrolyte diffusion problem.

Figure 5.4 shows the evolution of the concentration distribution with time. The concentration is initially zero across the cell. As time moves on the concentration in the anode increases and the cathode decreases. The results are plotted as differences in concentration relative to $c(0, t)$, so the distribution is always negative. It is clear that the zero-flux boundary conditions are enforced at $x = 0$ and $x = L$. At the junction between the two domains ($x = L/2$), the concentration and flux are continuous. The slope of the concentration distribution has a slight kink at $x = L/2$ associated with the change in diffusion coefficient.

Discretization Efficiency

In this section we evaluate the numerical efficiency of the various solution methods that were presented in Chapter 4 relative to an exact or high-order model for step inputs. For model-based

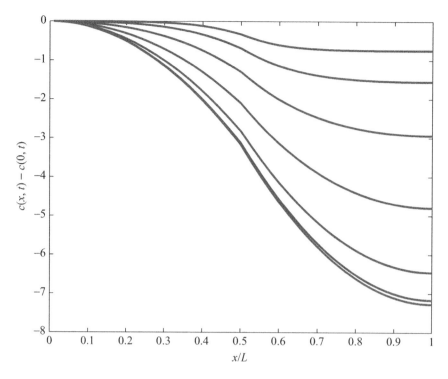

Figure 5.4 Concentration distribution $c(x, t) - c(0, t)$ in mol/m^3 for discharge step response in the electrolyte diffusion problem: analytical solution with 26-term approximation at $t = 1, 2, 4, 8, 16,$ and 32 s and exact steady-state solution (bottom curve)

estimation and control we need low-order models that accurately predict the time response of the battery. The exact, transcendental transfer function or a high-order approximation is used as the baseline for comparison. To implement an analytical model it must be truncated to produce a finite-order approximation. The other solution techniques automatically generate finite-order models with the order defined to be the number of integrators in the model. For different solution techniques, the order required to achieve a certain error determines the numerical efficiency of the model. This would be the model order (number of states) required for a truncated model of the desired accuracy. The step response is calculated at a sample rate of 10 Hz for 200 s or just over 2 min. This sample rate is typical of what is required for HEV models and control systems. A sample length of 200 s captures most of the low -speed dynamics.

The error between the exact $\mathbf{y}(t)$ and discretized and reduced-order $\hat{\mathbf{y}}(t)$ models is $\tilde{\mathbf{y}}(t) = \mathbf{y}(t) - \hat{\mathbf{y}}(t)$. The output vector $\mathbf{y}(t)$ can be chosen to either be the voltage (the only choice in comparing the model with experimental data) or internal variables (e.g., the ion concentration inside the cell). One must be careful to scale the elements of $\mathbf{y}(t)$ so that they are similar in magnitude. Otherwise, small elements that may have large relative errors will be swamped out by large elements with small relative errors. In practice, the response is simulated at a fixed sample frequency f_s and for a given time period $t \in [0, t_{\text{final}}]$, corresponding to N_s samples, where $N_s = t_{\text{final}} fs + 1$.

There are many possible error metrics that convert the error time distribution $\tilde{y}(t)$ for $t \in [0, t_{\text{final}}]$ to a positive number that measures the overall size of the error. The L_2 norm is

$$\|\tilde{\mathbf{y}}(t)\|_2 = \sqrt{\sum_{n=1}^{N_s} \left| \tilde{\mathbf{y}}\left(\frac{n}{f_s}\right) \right|_2^2}, \tag{5.8}$$

where $|\mathbf{y}|_2$ is the standard Euclidean norm. This norm penalizes small errors that persist over a long period of time but ignores large errors that appear for only a few samples. The L_∞ norm

$$\|\tilde{\mathbf{y}}(t)\|_\infty = \max_{n\in(1,N_s)} \left(\left| \tilde{\mathbf{y}}\left(\frac{n}{f_s}\right) \right|_\infty \right), \tag{5.9}$$

measures the worst-case error over the time period where

$$|\mathbf{y}|_\infty = \max_{i\in(1,N)} (|y_i|).$$

The L_2 and L_∞ metrics can be divided by $\|\mathbf{y}(t)\|_2$ and $\|\mathbf{y}(t)\|_\infty$ to produce error metrics as percentages of the response.

Figure 5.5 shows a typical discretization efficiency result for the two-domain electrolyte diffusion problem. In this plot the analytical model is truncated at different order N and the

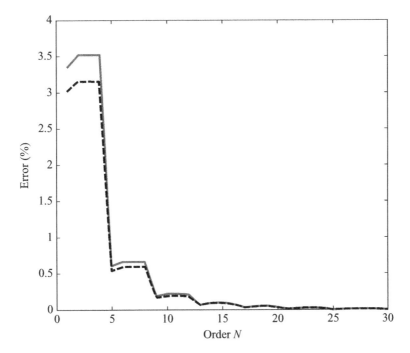

Figure 5.5 Step-response error metrics versus approximation order for the analytical solution of the two-domain electrolyte diffusion problem: L_2 norm $\|\tilde{\mathbf{y}}(t)\|_2$ (solid) and L_∞ norm $\|\tilde{\mathbf{y}}(t)\|_\infty$ (dashed)

Table 5.2 Approximation order required for electrolyte diffusion problem

	Step response				Frequency response			
	L_2		L_∞		L_2		L_∞	
Method[a]	0.5%	1%	0.5%	1%	0.5%	1%	0.5%	1%
PAM	1	1	2	2	3	3	3	3
IMA	4	3	4	4	4	3	4	3
RM	6	4	6	4	6	4	6	4
AM	9	5	9	5	9	5	9	5
FEM	10	10	10	12	10	12	10	14
FDM	27	15	27	15	27	15	27	15

[a]AM: analytical method; FDM: finite-difference method; FEM: finite-element method; IMA: integral method approximation; PAM: Padé approximation method; RM: Ritz method.

corresponding L_2 and L_∞ norms are calculated for the step response. The step-response plot for $N = 26$ is shown in Figure 5.3. The error is presented as a percentage of the L_2 or L_∞ norm of the response. Even for the lowest order analytical models ($N < 5$), the error is just over 3%. This may be sufficiently small for some applications. As expected, the accuracy generally increases (error metric decreases) with increasing model order. To achieve less than 1% error requires $N \geq 5$ and 0.5% error requires $N \geq 9$. The L_2 and L_∞ metrics follow the same trend with the L_∞ error slightly less than the L_2 error.

Table 5.2 summarizes the discretization efficiency of the six methods introduced in Chapter 4. The model orders required for 1% and 0.5% step-response errors are tabulated. The analytical method that was discussed in the previous paragraph lies in the middle of the pack. The Padé approximation method is the most efficient, requiring only $N = 2$ for 0.5% error for both L_2 and L_∞ norms. The IMA requires $N = 4$ to achieve the same results. Next is the Ritz method ($N = 6$), followed by the analytical method. The FEM at $N = 12$ and the FDM at $N = 27$ are the least efficient discretization methods.

Efficiency is one of several metrics that one can use to choose a discretization method. Although the FDM is not very efficient, it is the easiest method to apply in practice. Simplicity of model derivation can be helpful in reducing development costs and a model-order reduction algorithm could be used to reduce the order for implementation. Often, the extra time and effort required to develop an efficient model pays back dividends in the simulation, analysis, estimation, and control of battery systems. IMA and Ritz provide a nice trade-off between efficiency and simplicity of model development.

5.1.2 DST Cycle Response of the Pb–Acid Electrode

The DST, introduced in Chapter 1 (see Figure 1.3), is an example of the cycles that are used to test batteries for performance and life. The current input cycles between positive (charge) and negative (discharge) corresponding to a typical driving cycle and HEV mode. Battery systems engineering often requires the simulation of battery response to driving cycles like DST. In this section, the Pb–acid Ritz model developed in Chapter 4 is simulated with the DST cycle

Table 5.3 Pb–acid electrode model parameters

Parameter	Value
a_1	1.96×10^{-5} cm^2/s
a_2	0.256 /(mol cm^3 s)
a_3	987 A/(S cm^2)
a_4	3.29 cm^3/mol
a_5	0.0415 cm^2/s
a_6	5.40×10^{-7} cm^2/s

using the model parameters shown in Table 5.3 typical of a "benchmark" case that has been widely studied in the literature [36]. A Ritz model of order $N = 32$ was found to match the analytical concentration and overpotential frequency response very well from DC to 10 Hz.

Figure 5.6 shows the Pb-Acid electrode model time response to the DST current input. The current cycles between discharge and charge with a maximum charge rate of 1 A/cm^2 and discharge rate of -2 A/cm^2. The cycle has more discharge than charge so the concentration steadily decreases as shown in the top subplot. The voltage does not fade, however, because

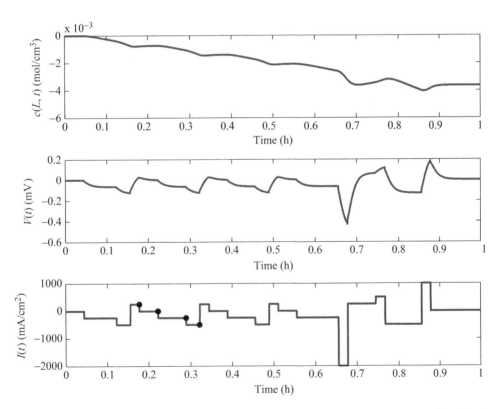

Figure 5.6 DST time response of the Pb–acid electrode model: (a) separator concentration $c(L,t)$; (b) voltage $V(t)$; (c) current $I(t)$

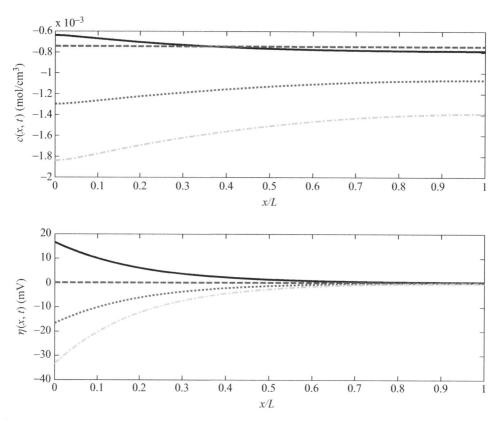

Figure 5.7 DST time response of the Pb–acid electrode model: (a) concentration distribution $c(x, t)$ and (b) overpotential distribution $\eta(x, t)$ at $t = 0.1778$ (solid), 0.2222 (dashed), 0.2889 (dotted), and 0.3217 h (dash–dotted line) (filled circles in Figure 5.6)

the output of the Pb electrode simulated here depends only on overpotential which does not decrease with decreasing acid concentration. In the PbO_2 electrode the acid concentration contributes to output voltage. For this cell the voltage response is associated with the resistive processes inside the electrode. Figure 5.7 shows the concentration (top) and overpotential (bottom) distributions at specific points in time. The time points correspond to the charge to discharge input between $t = 0.18$ and 0.32 h. The concentration distribution shows zero slope at the two boundaries, matching the prescribed boundary conditions. The overpotential distribution has a large slope at $x = 0$ and is almost zero at $x = L$, due to the magnitudes of the a_5 and a_6 coefficients. The slope sign at $x = 0$ clearly changes when the current changes direction, again in agreement with the overpotential boundary condition at $x = 0$.

5.2 Frequency Response

The frequency response of a battery reveals its dynamic characteristics over a wide frequency range. The relationships between the system parameters and the magnitude and phase of the response to sinusoidal inputs provides important insights into the dynamic behavior of the battery. For systems engineers, Bode plots of the frequency response can be used to understand the response, improve dynamic models, identify phenomena, and promote understanding of the battery system.

5.2.1   Electrochemical Impedance Spectroscopy

Electrochemists have long used the frequency response to experimentally study the battery characteristics with EIS. The EIS plot is the same as the Nyquist plot of the impedance transfer function $V(s)/I(s) = Z(s)$ except that the imaginary axis is typically shown with the negative direction pointing upward. The impedance is $Z(i\omega) = Z'(\omega) + iZ''(\omega)$, where Z' and Z'' are the real and imaginary parts of the impedance. Further information on EIS can be found in [37, 38].

Figure 5.8 shows the EIS results for a typical Li-ion battery. At low frequency (0.01 Hz) the Nyquist plot lies along a line with a slope of 45°, so the phase is nearly constant over this low-frequency range around 45°. A constant phase that persists over a wide frequency range and is not a multiple of 90° is foreign to most systems engineers, who are used to transfer functions resulting from linear ODEs. The underlying PDE models of battery cells cause some non-standard behavior, including phase asymptotes at multiples of 45°. At 0.79 Hz, the frequency response deviates from the 45° line and creates a semicircular arc that crosses zero at 158 Hz. As frequency increases further, the Nyquist plot asymptotically approaches ∞i with a fixed Z' and $Z'' \to \infty$. Most applications operate in the frequency range of DC to 10 Hz, so the low- and mid-frequency arcs are of most interest.

One can approximate the frequency response of a battery cell using an equivalent circuit like the one shown in Figure 5.9. At high frequencies, the inductance L dominates the response because it has impedance $Z_l(\omega) = i\omega L$. The ohmic and charge transfer resistances, R_0 and R_{ct}, have real impedance equal to their resistance and, therefore, act over the entire frequency range. The capacitor has impedance $Z_c(\omega) = -i/\omega C_{dl}$ that becomes large in magnitude and

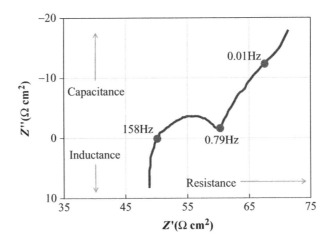

Figure 5.8   Example EIS experimental data for an Li-ion battery

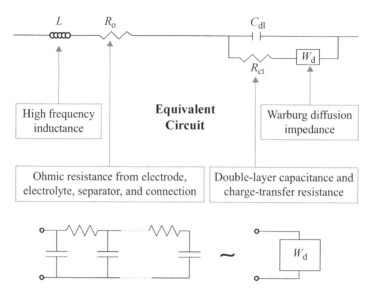

Figure 5.9 Example electrode equivalent circuit with high-frequency inductance, ohmic resistance, double-layer capacitance, charge-transfer resistance, and Warburg diffusion impedance. An equivalent-circuit model of the Warburg diffusion impedance is shown at the bottom

negative at low frequency. The Warburg diffusion resistance is designed to simulate the diffusion through particles and/or the electrolyte of the cell. Recall that the analytical transfer functions we derived for electrolyte and particle diffusion models were functions of the $\sqrt{s}$. Finite-dimensional models can only approximate the fractional derivatives in these distributed (infinite-dimensional) models. The Warburg diffusion impedance is expressed as

$$W_d(s) = \frac{\sigma}{s^n}, \tag{5.10}$$

where σ is a variable gain. The exponent n is often allowed to deviate from 0.5 to better fit the experimental data and reflect the complexity of different size particles and porous jelly roll structures in the cell. If $n = 1/2$, however, the Warburg impedance

$$W_d(i\omega) = \frac{\sigma}{\sqrt{i\omega}} = \frac{\sigma}{\sqrt{\omega}} \left[\cos\left(\frac{\pi}{4}\right) - i\sin\left(\frac{\pi}{4}\right) \right], \tag{5.11}$$

so $|W_d(i\omega)| = \sigma/\sqrt{\omega}$ and

$$\angle W_d(i\omega) = \tan^{-1}\left[\tan\left(-\frac{\pi}{4}\right)\right] = -\frac{\pi}{4} \tag{5.12}$$

or $-45°$.

The Warburg diffusion impedance is negative (or capacitive) and becomes large at low frequencies. It decreases less quickly with decreasing frequency than capacitance does, so it dominates the response as $\omega \to 0$. The Warburg impedance can be approximated by the

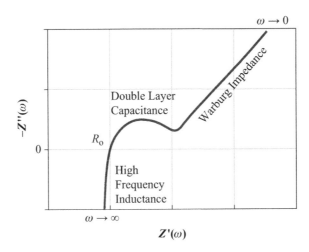

Figure 5.10   Nyquist plot of example equivalent circuit in Figure 5.9

R–C ladder circuit shown in Figure 5.9. To approximate the equivalent circuit with a finite-dimensional model one must select the number of capacitors (and hence model order) that approximates the distributed Warburg impedance with sufficient accuracy in the operating frequency range.

Figure 5.10 shows that the equivalent-circuit model in Figure 5.9 produces a Nyquist plot that closely mirrors the experimental result in Figure 5.8. The Warburg impedance provides the 45° line at low frequency. The double-layer capacitance in parallel with the charge transfer resistance provide the mid-frequency arc. As $\omega \to \infty$ the inductance dominates and the impedance approaches $R_0 + \infty i$.

Electrolyte Diffusion

Figure 5.11 shows the frequency response of the electrolyte diffusion model. The overall shape of the concentration frequency response is that of a low-pass filter. The concentration has a steady-state response at low frequency and rolls off at high frequency. The corner frequency is around 3×10^{-2} Hz. The exact solution is calculated by substituting $s = i\omega$ into the transcendental transfer function (4.63) and calculating the associated gain and phase. The exact solution is hidden behind the analytical solution with 26 modes. The analytical solution converges as the number of terms in the truncated series increases from 2 to 4 to 26. Again, only a few modes are required to accurately capture the frequency response across the bandwidth of interest.

5.2.2   Discretization Efficiency

The accuracy with which a model predicts the frequency response depends on the definition of an appropriate error metric. We use the same error vector $\tilde{Y}(i\omega)$ except that it is now in the Laplace domain and evaluated at $s = i\omega$. The frequency domain is divided into N_s samples

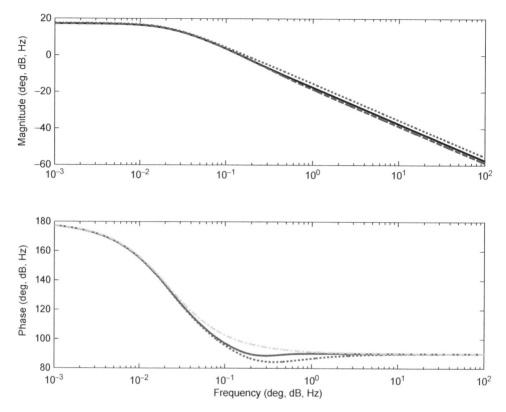

Figure 5.11 Bode plot of electrolyte diffusion: exact solution (solid black) and analytical solution with 26- (dashed), 4- (dash–dotted) and 2-term (dotted) approximations; output concentration $(C(L, i\omega) - C(0, i\omega))/I(i\omega)$

using a log spacing from f_{min} to f_s (in Matlab: f = logspace(log10(fmin), log10(fs), Ns)). The L_2 and L_∞ norms convert the $\tilde{\mathbf{Y}}(i\omega_n)$ complex frequency distribution to a positive number that measures the overall size of the error. The L_2 frequency norm is

$$\|\tilde{\mathbf{Y}}(i\omega)\|_2 = \sqrt{\sum_{n=1}^{N_s} |\tilde{\mathbf{Y}}(i\omega_n)|_2^2} \qquad (5.13)$$

This norm penalizes small errors that persist over the entire frequency range but ignores large errors that appear for only a few samples. The L_∞ norm

$$\|\tilde{\mathbf{Y}}(i\omega)\|_\infty = \max_{n \in (1, N_s)} \left(|\tilde{\mathbf{Y}}(i\omega_n)|_\infty \right), \qquad (5.14)$$

measures the worst-case error over the frequency range of interest. The L_2 and L_∞ metrics can be divided by $\|\mathbf{Y}(i\omega)\|_2$ and $\|\mathbf{Y}(i\omega)\|_\infty$ to produce error metrics as percentages of the response.

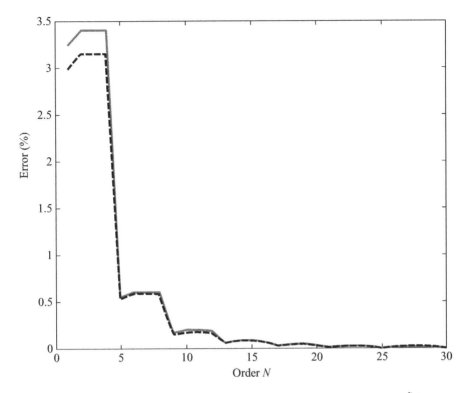

Figure 5.12 Frequency-response error metrics versus approximation order: L_2 norm $\|\tilde{Y}(i\omega)\|_2$ (solid) and L_∞ norm $\|\tilde{Y}(i\omega)\|_\infty$ (dashed)

In this section we evaluate the numerical efficiency of the various solution methods that were presented in Chapter 4 relative to an exact or high-order frequency response. The frequency response is calculated from $f_{min} = 0.005$ Hz to $f_s = 10$ Hz with $N_s = 2000$ to be consistent with the error metrics used in the step response analysis.

Figure 5.12 shows the discretization efficiency results for the analytical method. The results are presented as a percentage of the L_2 or L_∞ norm of the response and look very similar to the step response results shown in Figure 5.5. The two metrics also track each other with the L_∞ norm slightly less than L_2. For less than 1% and 0.5% error, the analytical method requires $N = 5$ and 9, respectively. The 1% and 0.5% L_2 and L_∞ error results for the six discretization methods are summarized in Table 5.2. The frequency-response results closely mirror those for the step response.

Li-Ion Particle

The Li-ion particle transfer function given in Equation (4.84) provides an interesting testbed for evaluating discretization techniques in the frequency domain. Figure 5.13 shows the frequency response for typical Li-ion particle parameters ($D = 2 \times 10^{-12}$ cm^2/s, $a_s = 17\,400$ cm^2/cm^3, and $R = 1$ μm). The pole at the origin is removed by subtracting the integrator from the

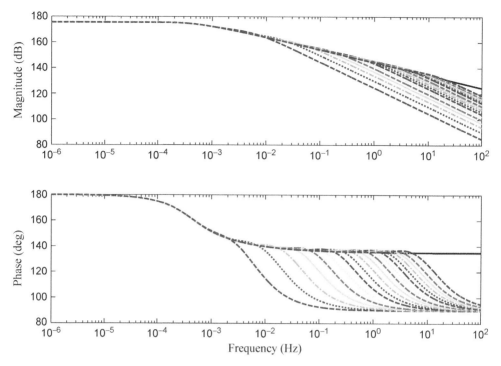

Figure 5.13 Bode plot of Li-ion particle transfer function (solid black) with Padé approximations varying from second to 19th order

transfer function to simplify the problem. At low frequency the transfer function approaches a negative constant. At high frequency the magnitude decays to zero at -10 dB/dec and the phase approaches $135°$. Polynomial transfer functions can only roll off at multiples of s, giving 0 dB/dec, ±20 dB/dec, ±40 dB/dec, ... slope and $0°$, $\pm90°$, $\pm180°$, ... phase asymptotes. Any polynomial approximation eventually has to go to a multiple of $90°$ as frequency goes to infinity. To match the transcendental transfer function over a wide frequency range, therefore, requires a relatively high-order approximation.

The second-order Padé approximation of this transfer function is derived in Equation (4.86). Figure 5.13 shows the second- to 19th-order Padé approximations of the Li-ion particle transfer function. The Padé approximation is based on matching the derivatives of the polynomial and transcendental transfer functions as $s \rightarrow 0$. All of the Padé approximations match the exact solution at low frequency ($< 10^{-2}$ Hz). The Padé approximations asymptote to $90°$ and -20 dB/dec, not the desired $135°$ and -10 dB/dec. As the approximation order increases, the frequency range that matches the exact solution increases in a fairly regular manner. Each additional term, however, only yields a small increase in bandwidth so to reach the desired 10 Hz with a high degree of accuracy requires a 19th-order model. Small deviations in phase and magnitude have little effect on the response at 10 Hz because the transfer function magnitude is quite small. A lower order Padé approximation may be adequate for this reason.

The applications of the Padé approximation are limited because it requires derivatives of the governing transfer function. For the spherical particle transfer function one can obtain

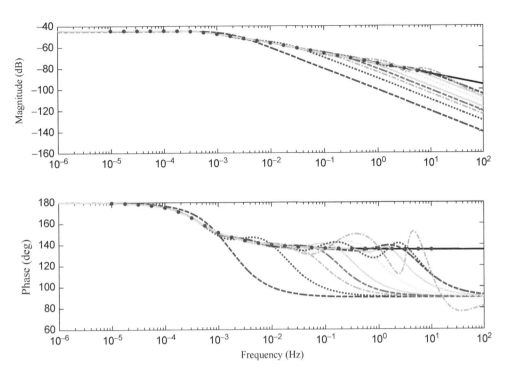

Figure 5.14 Bode plot of Li-ion particle transfer function (solid black) with system identification approximations varying from first to tenth order

analytical derivatives but they are quite difficult to derive, even using a symbolic math program like Maple. In general, it is only possible to generate derivatives for the simplest systems. In most cases, analytical derivatives will be difficult or impossible to obtain. Numerical derivatives are usually not suitable because the Padé approximation is very sensitive, especially at high order, and can produce inaccurate or even unstable approximations if the derivatives are not precise.

System identification is another method to generate low-order models from frequency-domain data. Analytical derivatives are not required, but the method is numerical with fewer guarantees on convergence. The quality of the approximation can be evaluated using frequency-domain metrics for different model orders and input parameters to the optimizer. With some effort a suitable approximation can often be found.

Figure 5.14 shows the system identification approximations varying from first to tenth order. The black line is the exact solution with circles superimposed to show the 25 points at which the error is calculated. Increasing the number of evaluation points to 50 or 100 has little effect on the results. The evaluation points range from $f_{min} = 10^{-5}$ Hz to the desired bandwidth of $f_{max} = 10$ Hz. Picking the low-frequency cutoff can have a significant impact on the identified transfer function. The least squares technique treats the errors from all the evaluation points the same. The evaluation points at high frequency have much smaller magnitude so they are not weighted as heavily. The evaluation points should balance the high and low frequency ranges of interest with more points biased toward high frequency to maintain accuracy in that range. One can premultiply the cost function by a function that increases in magnitude with frequency to increase the weighting on higher frequency response. The residues are found from linear least squares and the Matlab function lsqnonlin.m from the optimization toolbox is used to find the poles. The Jacobian is used and the options are set to Tolfun = 6e-12 and TolX = 6e-9.

Random initial guesses are used to help find a global minimum. The guess for the first pole location is a random number within the range $-2\pi f_{min}$ to $-2\pi f_{max}$ using a distribution that is uniform in the log scale. The second guess is constrained between the first pole and the minimum $(-2\pi f_{max})$, again using a uniform, log-scale random distribution. Each additional pole guess is constrained to be within the previous guess and the minimum. If the initial guesses are too close together then there can be singularity problems within the optimizer. The distance between adjacent pole guesses is forced to be greater than a user-specified minimum. The minimum error model for each order is shown in Figure 5.14. Two hundred and fifty initial guesses are used for the results shown. Increasing the number of guesses does not significantly change the model accuracy, so the lower value of 250 is used to minimize computation time.

Figure 5.14 shows the results for the system identification algorithm. The first-order approximation minimizes the mean-squared-error between the 25 evaluation points and a single pole/residue term. The approximation matches well at low frequency, where the magnitude is high, and not well at high frequency, where the magnitude is low. The identification accuracy increases fairly regularly with increasing model order, but the seventh order oscillates through the frequency domain. This may be a local solution that just happens to be minimal for the 250 initial guesses or may actually be the global minimum. The best solution is 10th order and shown as a dashed line in Figure 5.14.

The real and imaginary parts of the spherical-particle frequency response can also be plotted following the Nyquist or EIS convention. For consistency, Figure 5.15 shows the frequency response on the complex plane using the EIS convention, where the negative imaginary axis is pointing upward. Unlike the previous figures, the integrator pole at the origin is retained. To compare with the EIS data, we need to relate the concentration of lithium ions on the surface of the particle with the voltage. There is typically a negative trend in voltage with increasing concentration, so the negative of the transfer function is plotted in Figure 5.15. The frequency increases from the top right to bottom left with the 0.1 Hz, 1 Hz, and 10 Hz points shown with a circle, square, and diamond, respectively. The low-frequency response has the typical Warburg impedance tail at 45°. There is a little bump in the response at around 5 Hz that could be construed as the start of a semicircular arc. The spherical particle model, however, does not include double-layer capacitance, contact resistance, or inductance, so the frequency response deviates from a typical cell EIS at higher frequencies. Inclusion of contact resistance would shift the frequency response to the right. Double-layer capacitance would produce an obvious semicircular arc. Inductance would cause the imaginary part of the EIS curve to dip below zero at high frequency. If we zoom out and look at even lower frequencies, the integrator dominates and the frequency response tends to infinity along the $-Z''$ axis (i.e., phase of $-90°$).

5.3 Model Order Reduction

In the model-based control of complex or large-scale systems, model order reduction enables low-order controllers that are easier to design, implement, and operate in real time. Model order reduction techniques have been developed that decompose a mathematical model into

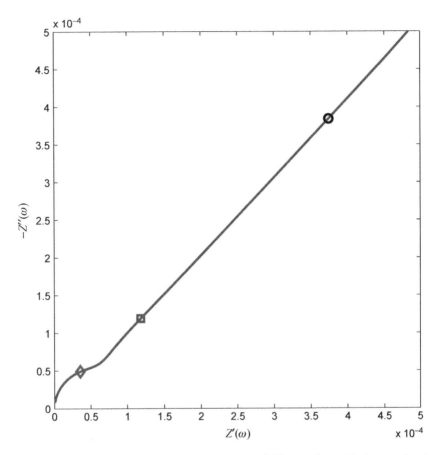

Figure 5.15 EIS plot of Li-ion particle transfer function (solid line) with specific frequencies shown at $f = 0.1$ Hz (circle), $f = 1$ Hz (square), and $f = 10$ Hz (diamond)

modes and sort those modes based on dominance (either magnitude or speed), controllability, or observability. Unimportant modes may be discarded or, in the case of fast modes, retained as static gains. Battery models typically involve infinite-dimensional PDEs, complicating the identification of an accurate low-order model. Detailed discussions of model order reduction techniques are provided in [39–42].

In this section, four model order reduction techniques are applied that produce low-order and accurate models. First, the model is truncated at the lowest N_{red} modes. Second, a standard model order reduction technique, based on the balanced realization algorithm in Matlab (balred.m) is used. The balred algorithm simply requires inputs of the full-order model and the desired order of the reduced-order model. The method discards the states with relatively small Hankel singular values while matching the DC gains between the full- and reduced-order models [43]. Third, model order is reduced by grouping, or lumping together, modes with similar eigenvalues [44]. This method takes advantage of the real eigenvalue spectrum of battery models. Finally, frequency-response curve fitting is introduced that can be used to convert a transcendental transfer function to a low-order rational polynomial transfer function.

5.3.1   Truncation Approach

Truncation is the simplest method of reducing model order. For SISO parabolic PDE systems that include most battery models, we can represent the transfer function as

$$G(s) = \sum_{n=1}^{\infty} \text{Res}_n s - \lambda_n \tag{5.15}$$

if the poles are real and not repeated. The residues are calculated from

$$\text{Res}_n = \lim_{s \to \lambda_n} (s - \lambda_n) G(s). \tag{5.16}$$

A truncated model stops the infinite series in Equation (5.15) at the first N terms. The eigenvalues λ_n are real for parabolic PDE systems and we order them so that $|\lambda_0| < |\lambda_1| < \cdots < |\lambda_{N-1}|$. The corresponding residues tend to have $|\text{Res}_0| > |\text{Res}_1| > \cdots |\text{Res}_{N-1}|$. To chose the truncation order N, we can either discard the modes that are faster than the sample frequency f_s, such that

$$|\lambda_{N-1}| \leq 2\pi f_s < |\lambda_n| \quad \text{for } n \geq N, \tag{5.17}$$

or discard the modes with small residues

$$|\text{Res}_{N-1}| > \text{Res}_{\min} > |\text{Res}_n| \quad \text{for } n \geq N. \tag{5.18}$$

Modes with small residues are also relatively uncontrollable and/or observable. Truncation can be an effective model-order reduction strategy for some battery models. Others, however, have many eigenvalues with similar residues in the frequency range of interest and another method of model order reduction is required.

5.3.2   Grouping Approach

In a grouping approach, we partition the frequency range of interest into N "bins" and lump together modes within each bin. Grouping indices $k_f \in \{0, 1, 2, \ldots n\}$ are arranged such that $0 = k_0 < k_1 < \ldots < k_N = n$. The grouped residue corresponding to bin $f \in \{1, 2, \ldots, N\}$ is

$$\overline{\text{Res}}_f = \sum_{k=k_{f-1}+1}^{k_f} \text{Res}_k, \tag{5.19}$$

with corresponding residue-weighted pole

$$\overline{p}_f = \frac{\displaystyle\sum_{k=k_{f-1}+1}^{k_f} p_k \text{Res}_k}{\overline{\text{Res}}_f}, \tag{5.20}$$

where the residues Res_k and poles p_k are calculated from the original transfer function that is to be reduced in order. Equation (5.20) places the grouped pole near the mode with dominant response and allows closely spaced modes with opposite sign residues to cancel one another. The grouping procedure yields a Nth-order transfer function

$$\frac{Y^*(x,s)}{U(s)} = \sum_{f=1}^{N} \frac{\overline{\mathrm{Res}}_f}{s - \overline{p}_f}. \tag{5.21}$$

An alternative, less accurate, grouping approach is to choose the grouped poles $\overline{p}_k$ a priori and then find the grouped residues $\overline{\mathrm{Res}}_k$ that most closely reproduce the grouping result in Equations (5.19) and (5.20). Battery models often have so many eigenvalues that as long as the grouped poles span the frequency range of interest their exact values are not so important. In effect, the modeler can place the poles of the *open-loop* system – a foreign concept for most systems engineers. Alternatively, the residues can be numerically optimized to minimize the error between the full- and reduced-order models. The advantage of choosing or placing the open-loop poles using this approach is that multiple transfer functions can share the same eigenvalues. This allows the development of lower order models for the battery system as a whole. It also allows models of different batteries at different temperatures and ages to share the same eigenvalues. In the state-variable model, changes due to aging, temperature, and so on then only appear in the output matrix, simplifying parameter estimation.

5.3.3 Frequency-Response Curve Fitting

The optimization approach described in Section 4.7.3 can also be applied to model order reduction. In Section 4.7.3 the objective is to optimize the match between the theoretical transcendental transfer function and a discrete polynomial transfer function. In model order reduction, the objective is to match the frequency response of a given, high-order polynomial transfer function with a lower order approximation. The same algorithms and approach can be applied, however.

5.3.4 Performance Comparison

To understand why model order reduction is often required and compare the various model-order reduction techniques, we use the spherical particle model as an example. This model has a relatively simple transfer function:

$$G(s) = \frac{\tanh(\Gamma(s))}{\tanh(\Gamma(s)) - \Gamma(s)}, \tag{5.22}$$

where $\Gamma(s) = R\sqrt{s/D}$. To simplify the analysis we substitute $s = -\gamma^2 D/R^2$ to produce

$$G(s) = \frac{\sin(\gamma)}{\sin(\gamma) - \gamma \cos(\gamma)}. \tag{5.23}$$

The eigenvalues are the roots of the denominator of $G(s)$ or

$$\tan(\gamma) - \gamma = 0. \tag{5.24}$$

Table 5.4   First five Li-ion particle eigenvalues

Truncation order	Exact	Approximate
1	0	0
2	4.49	4.71
3	7.72	7.85
4	10.9	11.0
5	14.1	14.1

Table 5.4 shows the first five roots of Equation (5.24). The eigenvalues are then calculated from $\lambda_n = -\gamma_n^2 D/R^2$. For a typical Li-ion particle, $R = 1.0\ \mu m$ and $D = 2.0 \times 10^{-12}\ cm^2/s$, so $R^2/D = 5000$ s. After the first eigenvalue at the origin, the second eigenvalue equals 6.4×10^{-4} Hz. The lowest frequencies of the Li-ion particle are very slow indeed. For higher order eigenvalues, an approximation can be used because $\gamma \to \infty$. From Equation (5.24), $\tan(\gamma) \to \infty$ or

$$\gamma_a = \frac{\pi}{2}(2n-1) \quad \text{for } n > 0. \tag{5.25}$$

The approximate solution quickly converges to the exact value with increasing n, as shown in Table 5.4 with an error of only 0.5% at $n = 5$.

Figure 5.16 shows the eigenvalues (in hertz) of the Li-ion particle as a function of truncation order for the truncation model order-reduction approach. In the log scale, the eigenvalues increase quickly for low n and then more slowly as $n \to \infty$. Application of the criteria in Equation (5.17) at $f_s = 10$ Hz would require almost 200 modes. The residues of $G(s)$ are

$$\text{Res}_1 = -\frac{3\,D}{R} \quad \text{and} \quad \text{Res}_n = -\frac{2\,D}{R} \quad \text{for } n > 1. \tag{5.26}$$

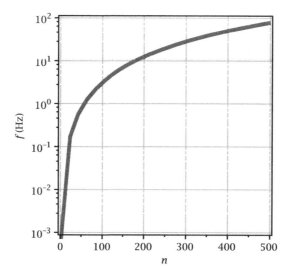

Figure 5.16   Li-ion particle eigenvalues versus truncation order

Thus, we cannot apply the criteria in Equation (5.18) because the residues do not decrease with increasing truncation order.

The effect of each pole on the system response is relatively small because for every pole there is a zero nearby. The zeros of the spherical transfer function are at $\sin(\gamma) = 0$ or $\gamma_m = (n-1)\pi$. As $\gamma \to \infty$, the poles and zeros alternate with $\gamma_m - \gamma_n = \pi/2$. This almost cancelation means that each pole has small but roughly equal contribution to the response.

The Li-ion particle model demonstrates that batteries have a wide spectrum of dynamics from time constants on the order of an hour to faster than any desired sampling frequency. Truncation does not provide efficient low-order models in some cases because there are too many modes below the cutoff frequency and the residues do not decrease with order. Thus, neither of the truncation criteria can be met with a low-order model.

The results for frequency-response curve fitting of the spherical particle model are presented for comparison in Section 5.2.2. Figure 5.14 shows that a reasonably good match in magnitude and phase up to 10 Hz can be achieved with a model order of less than 10. This is more than an order of magnitude smaller than what can be obtained using the truncation method.

Figures 5.17 and 5.18 show the results from balred.m and the grouping method. The grouping method always results in real residues and poles. In this example, balred also gives a reduced-order model with real poles and residues. Figure 5.17 plots the eigenvalues versus residues for the analytical, balred, and grouping methods. As we have already seen, the analytical residues are constant at around 1×10^{-5} and the analytical poles become more

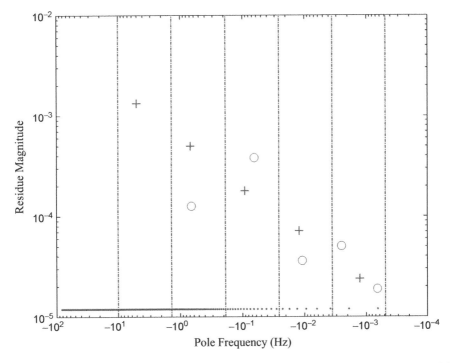

Figure 5.17   Li-ion particle model residues versus poles: analytical (•), balred (○), and grouped (+)

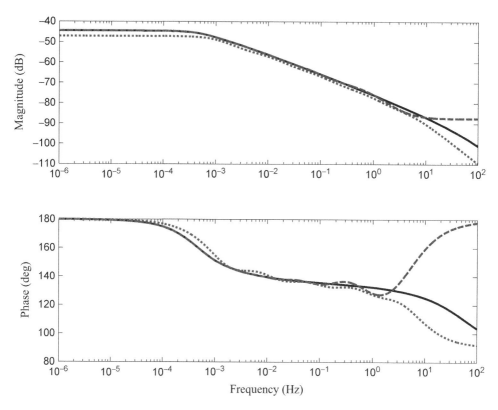

Figure 5.18 Li-ion particle model frequency response: analytical truncated to 512 modes (solid), balred (dashed), and grouped (dotted)

closely spaced with increasing frequency on the log-scale plot. The balred poles and residues are marked with ∘. The model reduction order is five and the five balred poles lie between 6×10^{-4} and 1 Hz. The balred residues generally increase in magnitude with increasing frequency. The five bins for the grouping approach are delineated by dash–dotted vertical lines. Within each bin the grouped pole is shown with a +. The bins are chosen based on the desired bandwidth (10 Hz) and the minimum analytical pole. The grouped poles are typically near the center of the bin with residues that increase in magnitude with increasing frequency.

Figure 5.18 compares the analytical, balred, and grouped frequency responses for $N = 5$. The analytical response is for the truncated model with 512 poles. This matches the analytical response from the transcendental transfer function shown in Figure 5.14 below the desired bandwidth of 10 Hz. Of course, the reduced -order models approximate the discretized 512th-order discrete model not the exact result in Figure 5.14. The fifth-order balred model closely matches the analytical response over the frequency range of interest with two orders of magnitude less states. The grouped results are not as accurate, with the magnitude coming below the analytical results over the entire frequency range and most significantly at low frequency. Phase, on the other hand, matches fairly well below 10 Hz.

Problems

5.1 The single-domain diffusion problem is governed by the PDE

$$\frac{\partial c}{\partial t} = a_1 \frac{\partial^2 c}{\partial x^2} \quad \text{for } x \in (0, L), \tag{A.5.1}$$

where $c(x, t)$ is concentration. The boundary conditions are

$$\frac{\partial c}{\partial x}\bigg|_{x=0} = 0 \tag{A.5.2}$$

and

$$\frac{\partial c}{\partial x}\bigg|_{x=L} = a_3 I(t), \tag{A.5.3}$$

where $I(t)$ is a current input. The constants $a_1 = D/\varepsilon$ and $a_3 = 1/(AF)$, where D is the diffusion coefficient, ε is the electrolyte phase volume fraction, A is electrode plate area, and F is the Faraday constant with the numerical values given in Table 5.1. Analytical, modal, IMA, Ritz, FEM, and FDM solutions are developed in Chapter 4. Use these solutions to compute and plot the frequency response and step response for this system as follows:
(a) Calculate and plot the transcendental (exact) frequency response of $C(L, s)/I(s)$ and compare with the plots for second-order (i) modal, (ii) IMA, (iii) Ritz, (iv) FEM, and (v) FDM approximations. Discuss the computational efficiencies of these methods relative to the efficiencies calculated for the coupled-domain problem shown in Table 5.2.
(b) For a unit impulse input $I(t)$, calculate the concentration distribution time response $c(x, t_i)$ with $t_i = 0, 1, 2, 8,$ and 16 s for the second-order (i) modal, (ii) IMA, (iii) Ritz, (iv) FEM, and (v) FDM models. Plot the distributions on the one plot for each method. Describe the time evolution of the response.

5.2 The coupled-domain diffusion problem is governed by the PDE

$$\varepsilon_m \frac{\partial c}{\partial t} = D_m \frac{\partial^2 c}{\partial x^2} + bI \quad \text{for } x \in (0, L/2), \tag{A.5.4}$$

$$\varepsilon_p \frac{\partial c}{\partial t} = D_p \frac{\partial^2 c}{\partial x^2} - bI \quad \text{for } x \in (L/2, L), \tag{A.5.5}$$

where

$$b = \frac{2(1 - t_0)}{FAL} \tag{A.5.6}$$

with the boundary conditions

$$\left.\frac{\partial c}{\partial x}\right|_{x=0} = \left.\frac{\partial c}{\partial x}\right|_{x=L} = 0. \tag{A.5.7}$$

The numerical values for the model parameters are given in Table 5.1. Analytical, Padé, IMA, Ritz, FEM, and FDM solutions are developed in Chapter 4. Use these solutions to compute and plot the frequency response and step response for this system as follows:

(a) Calculate and plot the transcendental (exact) frequency response of $C(L, s)/I(s)$ and compare with the plots for second-order (i) Padé, (ii) IMA, (iii) Ritz, (iv) FEM, and (v) FDM approximations. Discuss the computational efficiencies of these methods relative to the efficiencies calculated for the coupled-domain problem shown in Table 5.2.

(b) For a unit step input $I(t)$, calculate the concentration distribution time response $c(x, t_i)$ with $t_i = 0$, 1, 2, 8, and 16 s for the second-order (i) Padé, (ii) IMA, (iii) Ritz, (iv) FEM, and (v) FDM models. Plot the distributions and the steady state on the one plot for each method. Describe the time evolution of the response.

5.3 Calculate the frequency response $C(R, s)/J(s)$ of the second-order IMA discretization and the analytical transfer function derived for the spherical particle transfer function in Chapter 4 using the parameters from this chapter ($D = 2 \times 10^{-12}$ cm^2/s, $a_s = 17\,400$ cm^2/cm^3, and $R = 1$ μm). Plot the frequency responses of these two models on the same Bode plot. How well does the IMA result match the exact solution?

5.4 Calculate the analytical frequency response $C(R, s)/J(s)$ for the spherical and cylindrical particles derived in Chapter 4 using the same parameters ($D = 2 \times 10^{-12}$cm^2/s, $a_s = 17\,400$ cm^2/cm^3, and $R = 1$ μm). Plot the frequency responses of these two models on the same Bode plot. How do the frequency responses of the two particles differ?

5.5 Using the Ritz model developed in Chapter 4 with $N = 2$, simulate the voltage response of the Pb–acid electrode using the parameters from Table 5.3. Use the DST current input consisting of current pulses with

[duration (s), magnitude (A/cm^2)]
$$= [28, -0.25; 12, -0.5; 8, 0.25; 16, 0; 24, -0.25; 12, -0.5].$$

5.6 The analytical transfer function for solid-state diffusion in Li-ion particles is

$$\frac{L \tanh(L\sqrt{s/D})}{DAF[\tanh(L\sqrt{s/D}) - L\sqrt{s/D}]}.$$

Figure A.5.3 shows the FEM discretizations for $N = 20$, 40, 100, and 260 compared with the exact result using the analytical transfer function. Discuss the computational efficiency of the Ritz method for this problem.

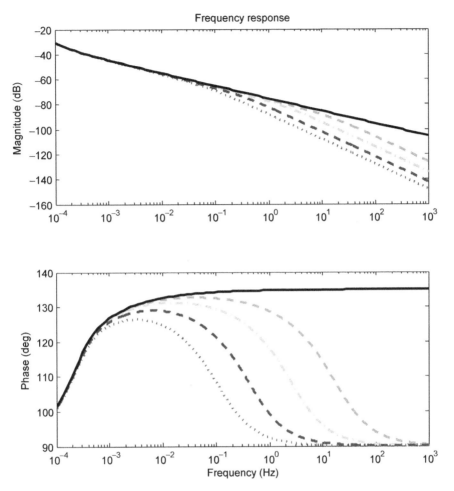

Figure A.5.3 Frequency response of the spherical particle transfer function: Ritz model with $N = 20$ (dotted), 40 (dashed), 100 (dash–dotted), and 260 (dashed) and analytical model (solid)

5.7 For solid-state diffusion in a cylindrical domain, the normalized transfer function from current input to surface concentration is

$$G(s) = \frac{I_0(\Gamma(s))}{\Gamma(s)I_1(\Gamma(s))},$$

where $\Gamma(s) = \sqrt{-s}$ and $I_0(s)$ and $I_1(s)$ are zeroth- and first-order Bessel functions of the first kind. Table A.5.1 shows the second- through fifth-order Padé approximations of $G(s)$. Plot the analytical frequency response and the four Padé approximations on

Table A.5.1 Padé approximations of diffusion in a cylindrical domain

Order	Padé approximation
2	$\dfrac{-2 - \frac{2}{5}s - \frac{3}{320}s^2}{s\left(1 + \frac{3}{40}s + \frac{1}{1920}s^2\right)}$
3	$\dfrac{-2 - \frac{3}{7}s - \frac{5}{336}s^2 - \frac{1}{10\,080}s^3}{s\left(1 + \frac{5}{56}s + \frac{1}{672}s^2 + \frac{1}{322\,560}s^3\right)}$
4	$\dfrac{-2 - \frac{4}{9}s - \frac{7}{384}s^2 - \frac{5}{24\,192}s^3 - \frac{5}{9\,289\,728}s^4}{s\left(1 + \frac{7}{72}s + \frac{5}{2304}s^2 + \frac{5}{387\,072}s^3 + \frac{1}{92\,897\,280}s^4\right)}$
5	$\dfrac{-2 - \frac{5}{11}s - \frac{9}{440}s^2 - \frac{7}{23\,760}s^3 - \frac{7}{4\,866\,048}s^4 - \frac{1}{567\,705\,600}s^5}{s\left(1 + \frac{9}{88}s + \frac{7}{2640}s^2 + \frac{7}{304\,128}s^3 + \frac{1}{16\,220\,160}s^4 + \frac{1}{40\,874\,803\,200}s^5\right)}$

the same Bode plot. Comment on the convergence and computational efficiency of the Padé approximations. What order of Padé approximation is sufficient for a 10 Hz bandwidth?

6

Battery System Models

From a systems perspective, batteries are essentially SISO with the input being current and the output being voltage. Obtaining a transfer-function or state-space model of a battery cell, however, is extremely complex because the systems engineer must formulate a complete set of governing PDEs, discretize the PDEs to ODEs, determine the system parameters, and specify the current input profile. The first step is typically undertaken with the help of an expert in electrochemistry and battery modeling unless the specific cell chemistry and design are well known and available in the literature, as is the case for the example Pb–acid, Ni–MH, and Li-ion cells studied in this chapter. The second step uses the methods developed in Chapter 4 to reduce the infinite-dimensional governing equations to a finite-order model with as few states as possible. Specification of the current input profile is usually straightforward, as either pulse charge/discharge, duty cycle, or sinusoidal inputs are typically applied.

Determination of the model parameters, however, is a non-trivial step. There are many independent parameters that are unknown in a given battery. The battery manufacturer may be unwilling to divulge the parameters or not know their values. In theory, all of the parameters can be determined from independent tests on a given cell. It is almost always the case that the battery system engineer must make educated guesses about many parameters and then tune the parameter values to match the model predicted and experimentally measured responses. In some cases a reference electrode can be inserted into the separator so that the voltage developed in the two electrodes can be individually measured. This provides an additional output that can be used to tune the model. Finally, the system parameters change with temperature and cycling, so the assumption of constant values may not be accurate.

Cell models can be extended to battery and pack models by considering the series and parallel cell connections that are used to raise the voltage and current, respectively. Cells in series can be modeled by simply multiplying the output voltage of a single cell by the number of cells in series. Cells in parallel can be modeled by an increase in the cell area. In practice, the cells in a given battery or pack are not identical and the imbalances developed in the pack are of interest to the systems engineer. In that case, multiple cell models with different parameters can be combined in series and parallel to more accurately represent the pack.

In this chapter we present complete, discretized, models of Pb–acid, Ni–MH, and Li-ion cells and simulate their responses to a variety of current inputs. Ritz, analytical, FEM, and IMA

Battery Systems Engineering, First Edition. Christopher D. Rahn and Chao-Yang Wang.
© 2013 John Wiley & Sons, Ltd. Published 2013 by John Wiley & Sons, Ltd.

discretization techniques are applied. The voltage time response to step charge and discharge currents and frequency response are simulated. In addition to the voltage output, the time evolution of the internal potential and concentration distributions are presented. The models are linear, so they may not accurately represent the response at high current rates and/or overcharge/undercharge conditions. They can be used for performance simulations, however, and provide the basis for the estimation and management algorithms developed in the next two chapters.

6.1 Lead–Acid Battery Model

Pb–acid batteries are one of the oldest rechargeable technologies, with over 150 years of history. They have been widely used in both industry and consumer products because of their high reliabilty, power density and efficiency, and relatively low cost compared with other battery chemistries [45, 46].

As we saw in Chapter 2, the main electrochemical reactions in the positive electrode of a Pb–acid battery [47] are

$$PbO_2 + HSO_4^- + 3H^+ + 2e^- \underset{charge}{\overset{discharge}{\rightleftharpoons}} PbSO_4 + 2H_2O. \tag{6.1}$$

In the negative electrode,

$$Pb + HSO_4^- \underset{charge}{\overset{discharge}{\rightleftharpoons}} PbSO_4 + H^+ + 2e^-. \tag{6.2}$$

It is interesting to note that during discharge acid is consumed in both the negative and positive electrodes to form $PbSO_4$. During charge, $PbSO_4$ is converted back into acid in both electrodes.

The Pb–acid cell shown in Figure 6.1 consists of a lead dioxide electrode, separator, and lead electrode. Each electrode is a porous, electronically conductive matrix with the pores occupied by a sulfuric acid solution. Electrical currents are applied uniformly to the left and right vertical boundaries of the cell. Typical Pb–acid cells have a very large aspect ratio of height to thickness; that is, the height is hundreds of millimeters and the thickness is a few millimeters. In this model, we assume uniform response in the vertical direction and develop a one-dimensional model across the width of the cell. More details on model development for Pb–acid batteries can be found in [36, 48, 49].

6.1.1 Governing Equations

In this section we assemble the governing equations from Chapter 3 for a linear Pb–acid battery model. The spatial domain $x \in (0, L)$ divides into three domains: positive electrode for $x \in (0, L_1)$, separator for $x \in (L_1, L_2)$, and negative electrode for $x \in (L_2, L)$. The positive and negative electrodes are porous and saturated with electrolyte. The separator is also saturated with electrolyte but insulates the flow of electrons between the electrodes. The ions in the binary electrolyte migrate and diffuse through the electrodes and separator. Electrons flow through the external circuit.

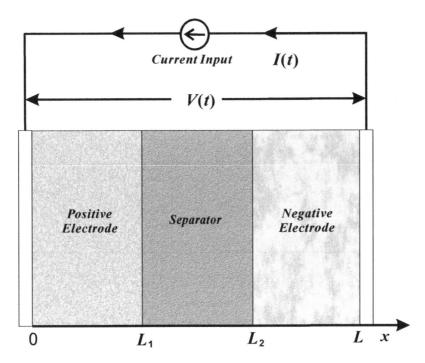

Figure 6.1   Pb–acid cell model

Several assumptions are made to simplify the model: (i) as discussed above, the variables are considered to be uniformly distributed or equal within cross-sections parallel to the current collectors; (ii) the parameters can take on different values in the positive electrode, separator, and negative electrode but are assumed to be constant within each of these domains; (iii) solid-phase potential ϕ_s is considered to be uniform everywhere in each electrode owing to the high conductivity of the solid-phase material; and (iv) gassing and other side reactions are neglected.

The system variables are the overpotential η, current density j, solid-phase potential ϕ_s, electrolyte phase potential ϕ_e, and electrolyte concentration c. The electrolyte variables ϕ_e and c are defined over the domain $x \in [0, L]$. The solid-phase variables η, j, and ϕ_s are defined over the domain $x \in [0, L_1] \cup [L_2, L]$ and are zero for $x \in (L_1, L_2)$. Assumption (i) means that the system variables are only functions of x and t.

Conservation of species in the electrolyte is governed by

$$\varepsilon \frac{\partial c}{\partial t} = D^{\text{eff}} \frac{\partial^2 c}{\partial x^2} + \frac{a_2}{2F} a j, \tag{6.3}$$

where $a_2 = 3 - 2t_+$ in the positive electrode and $a_2 = 1 - 2t_+$ in the negative electrode and the interfacial area a equals a_c during charge and a_d during discharge. The various parameters used in the model (e.g., D^{eff}, t_+, a_c, a_d, and F) are defined in Tables 6.1 and 6.2.

Table 6.1 Fixed system parameters

Quantity	Value
A: cell cross-section area (cm^2)	251.61
L_1: coordinate of boundary between positive electrode and separator (cm)	0.159
L_2: coordinate of boundary between negative electrode and separator (cm)	0.318
L: thickness of a cell (cm)	0.477
c_{ref}: reference H$^+$ concentration (mol/cm^3)	4.9×10^{-3}
Concentration of H$^+$ at fully charged state (mol/cm^3)	6.0×10^{-3}
R: universal gas constant (J/(mol K))	8.3143
F: Faraday constant (C/mol)	96 485
t_+: transference number of H$^+$	0.72
$\overline{U}_{PbO_2}$: setpoint open-circuit potential at 70% SOC (V)	1.8779
$\tilde{U}_{PbO_2}$: setpoint open-circuit voltage slope at 70% SOC (V/mol cm^3)	50.9
C_{dl}: specific capacitance of the double layer (C/cm^3)	2.0×10^{-4}

Charge conservation in the electrolyte is

$$\kappa^{eff}\frac{\partial^2 \phi_e}{\partial x^2} + \kappa_d^{eff}\frac{\partial^2 c}{\partial x^2} + aj + i_{dl} = 0, \qquad (6.4)$$

where the double-layer current per volume $i_{dl} = a_{dl}C_{dl}[\partial(\phi_s - \phi_e)/\partial t]$. The Butler–Volmer equation is linearized as

$$j = \frac{R_a}{a}\eta, \qquad (6.5)$$

where

$$R_a = ai_0 \left(\frac{\bar{c}}{c_{ref}}\right)^\gamma \frac{(\alpha_a + \alpha_c)F}{RT}$$

Table 6.2 Spatially varying system parameters at SOC = 70%

Parameter	Positive electrode	Separator	Negative electrode
ε: volume fraction of electrolyte	0.6454	0.8556	0.4433
D^{eff}: diffusion coefficient (cm^2/s)	1.4776×10^{-5}	2.2553×10^{-5}	8.4100×10^{-6}
a_c: specific interfacial area in charging (cm^2/cm^3)	3.9519×10^{5}	—	3.6967×10^{4}
a_d: specific interfacial area in discharging (cm^2/cm^3)	3.4808×10^{4}	—	6.0335×10^{3}
κ^{eff}: effective ionic electrolyte conductivity (1/(Ω cm))	0.1906	0.0573	0.4698
κ_d^{eff}: effective diffusional electrolyte (Acm^2/mol)	0.0021	0.0053	6.4316×10^{-4}
γ: exponent in Butler–Volmer equation	0.3	—	0
i_0: exchange current density (A/cm^2)	3.19×10^{-7}	—	4.96×10^{-6}
α_a: anodic transfer coefficient	1.15	—	1.55
α_c: cathodic transfer coefficient	0.85	—	0.45
a_{dl}: specific surface area of double layer (cm^2/cm^3)	4.3×10^{5}	—	4.3×10^{4}

and overpotential

$$
\eta = \begin{cases} \phi_{sp} - \phi_e - U_{PbO_2} & \text{+ve electrode} \\ 0 & \text{separator} \\ \phi_{sm} - \phi_e & \text{-ve electrode} \end{cases} \tag{6.6}
$$

The linear approximation of the Butler–Volmer equation (6.5) introduces negligible error at small overpotential η when the input current is not large. The open-circuit voltage U_{PbO_2} is a nonlinear function of acid concentration c given by

$$
\begin{aligned}
U_{PbO_2} &= 1.9228 + 0.0641 \ln(R_m) + 0.0120 \ln^2(R_m) \\
&\quad + 0.0060 \ln^3(R_m) + 0.0012 \ln^4(R_m)
\end{aligned} \tag{6.7}
$$

and plotted in Figure 6.2, where $R_m = 1.003\,22c + 0.0355c^2 + 0.0022c^3 + 0.0002c^4$. Linearization of Equation (6.7) at average acid concentration $\bar{c}$ yields the approximation

$$
U_{PbO_2} = \overline{U}_{PbO_2} + \tilde{U}_{PbO_2}\tilde{c}, \tag{6.8}
$$

where $\overline{U}_{PbO_2} = U_{PbO_2}(\bar{c})$, $\tilde{U}_{PbO_2} = dU_{PbO_2}/dc\,|_{c=\bar{c}}$, $\tilde{c}$ is a small deviation from $\bar{c}$. Figure 6.2 shows that the linear approximation at $\overline{U}_{PbO_2} = 1.8744$ V and $\tilde{U}_{PbO_2} = 52.3$ V/mol cm^3 is

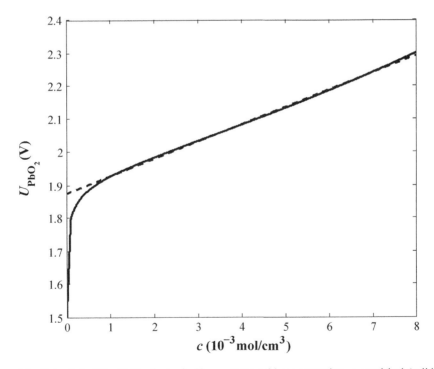

Figure 6.2 Potential of the PbO$_2$ electrode U_{PbO_2} versus acid concentration c: empirical (solid) and linear approximation (dashed)

very close to the actual U_{PbO_2} when $0.001 < c < 0.008$ mol/cm^3. This corresponds to a state of charge ranging from 17% to 100%.

Using assumption (iv), the integral of the reaction current plus double-layer current equals the input current in each electrode. For the positive electrode:

$$\int_0^{L_1} A \left[R_a(\phi_{sp} - \phi_e - U_{PbO_2}) + a_{dl} C_{dl} \frac{\partial(\phi_{sp} - \phi_e)}{\partial t} \right] dx = I(t), \qquad (6.9)$$

where assumption (ii) is used with $\phi_s(x, t) = \phi_{sp}(t)$. In the negative electrode:

$$\int_{L_2}^{L_3} A \left[R_a(\phi_{sm} - \phi_e) + a_{dl} C_{dl} \frac{\partial(\phi_{sm} - \phi_e)}{\partial t} \right] dx = -I(t), \qquad (6.10)$$

with $\phi_s(x, t)$ assumed equal to ϕ_{sm}. The voltage output is given by $V(t) = \phi_{sp} - \phi_{sm}$, where contact resistance has been neglected.

The three-domain battery model includes current collector and electrode–separator interface boundary conditions. At the current collectors ($x = 0$ and $x = L$):

$$\frac{\partial \phi_e}{\partial x} = 0, \quad \frac{\partial c}{\partial x} = 0. \qquad (6.11)$$

At the electrode–separator interfaces ($x = L_1$ and $x = L_2$):

$$\left(\kappa^{eff} \frac{\partial \phi_e}{\partial x} + \kappa_d^{eff} \frac{\partial c}{\partial x} \right)\bigg|_+ = \left(\kappa^{eff} \frac{\partial \phi_e}{\partial x} + \kappa_d^{eff} \frac{\partial c}{\partial x} \right)\bigg|_- ,$$

$$D^{eff} \frac{\partial c}{\partial x}\bigg|_+ = D^{eff} \frac{\partial c}{\partial x}\bigg|_- . \qquad (6.12)$$

Equations (6.3)–(6.12) constitute a complete set of equations for the linearized Pb–acid model. There are six independent equations to solve for the six unknowns ($\eta(x, t)$, $j(x, t)$, $\phi_{sp}(t)$, $\phi_{sm}(t)$, $\phi_e(x, t)$, and $c(x, t)$), given the input $I(t)$, initial conditions ($c(x, 0)$), and constant parameters in Tables 6.1 and 6.2. If we use dilute electrolyte theory then we can relate some of these parameters to others (e.g., D_+ and D_-), potentially reducing the number of independent parameters. The geometric parameters (L_1, L_2, L, and A) can often be obtained from the battery manufacturer. At the other end of the spectrum are the diffusion-related parameters (ε, D^{eff}, κ^{eff}, and κ_d^{eff}). These parameters are difficult to estimate within an order of magnitude. Thus, unless they are independently measured, they are prime candidates for tuning the system response.

6.1.2 Discretization using the Ritz Method

The Ritz method uses a set of admissible continuous functions as a basis for the L^2 space as described in Chapter 4. Here, we use a Fourier series expansion because the sinusoidal functions can be chosen to automatically satisfy most of the boundary conditions and the solution converges quickly with a low number of terms.

The acid concentration distribution is expanded in an Nth-order Fourier series:

$$c(x, t) = \sum_{m=0}^{N-1} \Psi_m(x) c_m(t), \tag{6.13}$$

where the admissible function $\Psi_m(x) = \cos\left(\frac{m\pi}{L}x\right)$. Electrolyte potential is expanded in an $N - 1$-order Fourier series:

$$\phi_e(x, t) = \sum_{m=1}^{N-1} \Psi_m(x) \phi_m(t). \tag{6.14}$$

The electrolyte potential expansion does not include the constant term that appears in the concentration distribution. This is equivalent to establishing that the average potential is zero across the cell. Alternatively, one can assign the negative current collector as ground. Equations (6.5) and (6.6) are substituted into Equation (6.4) to obtain the electrolyte potential equations for the three-domain model as

$$\begin{cases} \kappa^{\text{eff}} \dfrac{\partial^2 \phi_e}{\partial x^2} + \kappa_d^{\text{eff}} \dfrac{\partial^2 c}{\partial x^2} + R(\phi_{sp} - \phi_e - U_{PbO_2}) + a_{dl}C(\dot{\phi}_{sp} - \dot{\phi}_e) = 0 & x < L_1, \\[2ex] \kappa^{\text{eff}} \dfrac{\partial^2 \phi_e}{\partial x^2} + \kappa_d^{\text{eff}} \dfrac{\partial^2 c}{\partial x^2} = 0 & L_1 < x < L_2, \\[2ex] \kappa^{\text{eff}} \dfrac{\partial^2 \phi_e}{\partial x^2} + \kappa_d^{\text{eff}} \dfrac{\partial^2 c}{\partial x^2} + R(\phi_{sm} - \phi_e) + a_{dl}C(\dot{\phi}_{sm} - \dot{\phi}_e) = 0 & L_2 < x < L. \end{cases} \tag{6.15}$$

Substitution of the Ritz functions into Equation (6.15), multiplication by $\Psi_n(x)$, and integration across the domain produces

$$-\int_0^L \kappa^{\text{eff}} \Psi_{n'}(x) \left(\sum_{m=1}^{N-1} \Psi_m'(x) \phi_m(t) \right) dx - \int_0^L \kappa_d^{\text{eff}} \Psi_n'(x) \left(\sum_{m=1}^{N-1} \Psi_m'(x) c_m(t) \right) dx$$

$$+ \int_0^{L_1} R_a \Psi_n(x) \left(\phi_{sp} - \overline{U}_{PbO_2} - \tilde{U}_{PbO_2} \sum_{m=0}^{N-1} \Psi_m(x) c_m(t) - \sum_{m=1}^{N-1} \Psi_m(x) \phi_m(t) \right) dx$$

$$+ \int_0^{L_1} a_{dl} C_{dl} \Psi_n(x) \left(\dot{\phi}_{sp} - \sum_{m=1}^{N-1} \Psi_m(x) \dot{\phi}_m(t) \right) dx$$

$$+ \int_{L_2}^L R_a \Psi_n(x) \left(\phi_{sm} - \sum_{m=1}^{N-1} \Psi_m(x) \phi_m(t) \right) dx$$

$$+ \int_{L_2}^L a_{dl} C_{dl} \Psi_n(x) \left(\dot{\phi}_{sm} - \sum_{m=1}^{N-1} \Psi_m(x) \dot{\phi}_m(t) \right) dx = 0, \tag{6.16}$$

where integration by parts and the boundary conditions have been used. Conversion of Equation (6.16) into matrix form produces

$$\mathbf{M}_e\dot{\boldsymbol{\phi}}_e + \mathbf{M}_{es}\dot{\boldsymbol{\phi}}_s = \mathbf{K}_{ec}\mathbf{c} + \mathbf{K}_e\boldsymbol{\phi}_e + \mathbf{K}_{es}\boldsymbol{\phi}_s + \mathbf{B}_e\mathbf{u} \tag{6.17}$$

where

$$\mathbf{u} = \begin{bmatrix} I \\ \overline{U}_{PbO_2} \end{bmatrix}, \quad \boldsymbol{\phi}_e = \begin{bmatrix} \phi_{e,1}(t) \\ \vdots \\ \phi_{e,N-1}(t) \end{bmatrix}, \quad \boldsymbol{\phi}_s = \begin{bmatrix} \phi_{sp}(t) \\ \phi_{sm}(t) \end{bmatrix}, \text{ and } \mathbf{c} = \begin{bmatrix} c_0(t) \\ \vdots \\ c_{N-1}(t) \end{bmatrix}.$$

The matrices in Equation (6.17) are given in Table 6.3.

Similarly, the Fourier series expansions are substituted into Equations (6.9) and (6.10), premultiplied by $\Psi_n(x)$, and integrated to produce

$$\mathbf{M}_s\dot{\boldsymbol{\phi}}_s + \mathbf{M}_{se}\dot{\boldsymbol{\phi}}_e = \mathbf{K}_{sc}\mathbf{c} + \mathbf{K}_{se}\boldsymbol{\phi}_e + \mathbf{K}_s\boldsymbol{\phi}_s + \mathbf{B}_s\mathbf{u}. \tag{6.18}$$

Finally, Equation (6.3) can be processed to produce the Ritz approximation

$$\mathbf{M}_c\dot{\mathbf{c}} = \mathbf{K}_c\mathbf{c} + \mathbf{K}_{ce}\boldsymbol{\phi}_e + \mathbf{K}_{cs}\boldsymbol{\phi}_s + \mathbf{B}_c\mathbf{u} \tag{6.19}$$

Equations (6.17), (6.18), and (6.19) are combined in the state-space model

$$\mathbf{M}_1\dot{\mathbf{x}} = \mathbf{M}_2\mathbf{x} + \mathbf{M}_3\mathbf{u}, \tag{6.20}$$

where

$$\mathbf{M}_1 = \begin{bmatrix} \mathbf{M}_c & 0 & 0 \\ 0 & \mathbf{M}_e & \mathbf{M}_{es} \\ 0 & \mathbf{M}_{se} & \mathbf{M}_s \end{bmatrix}, \quad \mathbf{M}_2 = \begin{bmatrix} \mathbf{K}_c & \mathbf{K}_{ce} & \mathbf{K}_{cs} \\ \mathbf{K}_{ec} & \mathbf{K}_e & \mathbf{K}_{es} \\ \mathbf{K}_{sc} & \mathbf{K}_{se} & \mathbf{K}_s \end{bmatrix}, \quad \mathbf{x} = \begin{bmatrix} \mathbf{c} \\ \boldsymbol{\phi}_e \\ \boldsymbol{\phi}_s \end{bmatrix}, \text{ and } \mathbf{M}_3 = \begin{bmatrix} \mathbf{B}_c \\ \mathbf{B}_e \\ \mathbf{B}_s \end{bmatrix}.$$

In standard form, the Pb–acid state-space model can now be written as

$$\dot{\mathbf{x}} = \mathbf{A}_{ss}\mathbf{x} + \mathbf{B}_{ss}\mathbf{u}, \tag{6.21}$$

where $\mathbf{A}_{ss} = \mathbf{M}_1^{-1}\mathbf{M}_2$ and $\mathbf{B}_{ss} = \mathbf{M}_1^{-1}\mathbf{M}_3$.

For voltage output $y = V(t) = \phi_{sp} - \phi_{sm} + \overline{U}_{PbO_2}$ we have

$$y = \mathbf{C}_{ss}\mathbf{x} + \mathbf{D}_{ss}\mathbf{u}, \tag{6.22}$$

where $\mathbf{C}_{ss} = [0, \ldots, 0, 1, -1]$ and $\mathbf{D}_{ss} = [0, 1]$.

6.1.3   Numerical Convergence

One of the main advantages of the Ritz method is its fast convergence. This means that the Ritz approximation is able to quickly approach the exact solution as approximation order

Table 6.3 Ritz model matrix elements

Matrix	Integral form
$\mathbf{M}_e(n, m)$	$\int_0^{L_1} a_{dl}C_{dl}\Psi_n(x)\Psi_m(x)\,dx + \int_{L_2}^{L} a_{dl}C_{dl}\Psi_n(x)\Psi_m(x)\,dx$
$\mathbf{M}_{es}(n, 1)$	$-\int_0^{L_1} a_{dl}C_{dl}\Psi_n(x)\,dx$
$\mathbf{M}_{es}(n, 2)$	$-\int_{L_2}^{L} a_{dl}C_{dl}\Psi_n(x)\,dx$
$\mathbf{K}_e(n, m)$	$-\kappa^{eff}\int_0^{L}\Psi'_n(x)\Psi'_m(x)\,dx - \int_0^{L_1} R_a\Psi_n(x)\Psi_m(x)\,dx - \int_{L_2}^{L} R_a\Psi_n(x)\Psi_m(x)\,dx$
$\mathbf{K}_{ec}(n, m+1)$	$-\int_0^{L}\kappa_d^{eff}\Psi'_n(x)\Psi'_m(x)\,dx - \int_0^{L_1} R_a\widetilde{U}_{PbO_2}\Psi_n(x)\Psi_m(x)\,dx$
$\mathbf{K}_{es}(n, 1)$	$\int_0^{L_1} R_a\Psi_n(x)\,dx$
$\mathbf{K}_{es}(n, 2)$	$\int_{L_2}^{L} R_a\Psi_n(x)\,dx$
$\mathbf{B}_e(n, 1)$	0
$\mathbf{B}_e(n, 2)$	$-\int_0^{L_1} R_a\Psi_n(x)\,dx$
$\mathbf{M}_s$	$\begin{bmatrix} -a_{dl}C_{dl}L_1 & 0 \\ 0 & -a_{dl}C_{dl}(L-L_2) \end{bmatrix}$
$\mathbf{M}_{se}(1, n)$	$\int_0^{L_1} a_{dl}C_{dl}\Psi_n(x)\,dx$
$\mathbf{M}_{se}(2, n)$	$\int_{L_2}^{L} a_{dl}C_{dl}\Psi_n(x)\,dx$
$\mathbf{K}_{sc}(1, n+1)$	$-\int_0^{L_1} R_a\widetilde{U}_{PbO_2}\Psi_n(x)\,dx$
$\mathbf{K}_{sc}(2, n)$	0
$\mathbf{K}_{se}(1, n)$	$-\int_0^{L_1} R_a\Psi_n(x)\,dx$
$\mathbf{K}_{se}(2, n)$	$-\int_{L_2}^{L} R_a\Psi_n(x)\,dx$
$\mathbf{K}_s$	$\begin{bmatrix} R_aL_1 & 0 \\ 0 & R_a(L-L_2) \end{bmatrix}$
$\mathbf{B}_s$	$\begin{bmatrix} -\frac{1}{A} & -R_aL_1 \\ \frac{1}{A} & 0 \end{bmatrix}$
$\mathbf{M}_c(n+1, m+1)$	$\int_0^{L}\varepsilon\Psi_n(x)\Psi_m(x)\,dx$
$\mathbf{K}_c(n+1, m+1)$	$-\int_0^{L} D^{eff}\Psi'_n(x)\Psi'_m(x)\,dx - \widetilde{U}_{PbO_2}\int_0^{L_1}\frac{a_2R_a}{2F}\Psi_n(x)\Psi_m(x)\,dx$
$\mathbf{K}_{ce}(n+1, m)$	$-\int_0^{L_1}\frac{a_2R_a}{2F}\Psi_n(x)\Psi_m(x)\,dx - \int_{L_2}^{L}\frac{a_2R_a}{2F}\Psi_n(x)\Psi_m(x)\,dx$
$\mathbf{K}_{cs}(n+1, 1)$	$\int_0^{L_1}\frac{a_2R_a}{2F}\Psi_n(x)\,dx$
$\mathbf{K}_{cs}(n+1, 2)$	$\int_{L_2}^{L}\frac{a_2R_a}{2F}\Psi_n(x)\,dx$
$\mathbf{B}_c(n, 1)$	0
$\mathbf{B}_c(n+1, 2)$	$-\int_0^{L_1}\frac{a_2R_a}{2F}\Psi_n(x)\,dx$
$\int \Psi'_n(x)\Psi'_m(x)\,dx$	$\begin{cases} x & n=m=0 \\ \frac{L}{4n\pi}\sin\frac{2n\pi x}{L} + \frac{x}{2} & n=m\neq 0 \\ \frac{L}{2(n+m)\pi}\sin\frac{(n+m)\pi x}{L} + \frac{L}{2(n-m)\pi}\sin\frac{(n-m)\pi s}{L} & n\neq m \end{cases}$
$\int \Psi'_n(x)\Psi'_m(x)\,dx$	$\begin{cases} 0 & n \text{ or } m=0 \\ \frac{1}{2}\left(\frac{n\pi}{L}\right)^2\left(x-\frac{L}{2n\pi}\sin\frac{2n\pi x}{L}\right) & n=m\neq 0 \\ \frac{nm}{2}\left(\frac{\pi}{L}\right)^2\left[\frac{L}{(n-m)\pi}\sin\frac{(n-m)\pi x}{L} - \frac{L}{(n+m)\pi}\sin\frac{(n+m)\pi x}{L}\right] & n\neq m \end{cases}$

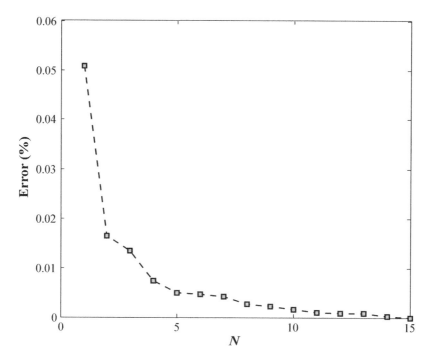

Figure 6.3 Convergence of the Ritz approximation: L_2 error versus model order

N increases. Figure 6.3 shows the L_2 error for a step input relative to a 15th-order Ritz approximation. At $N = 10$, the error is reduced to within 0.0015% of the $N = 15$ value.

6.1.4 Simulation Results

The parameters shown in Tables 6.1 and 6.2 and used in the simulations presented in this section come from a VRLA AGM battery with a capacity of 70 Ah with eight cells in parallel and six in series. The Ritz model is simulated for constant discharge and charge starting from 70% SOC with $N = 8$. Figure 6.4 shows both the discharge and charge responses at $0.1C$. For a linear model the transients are identical but flipped in sign for inputs of opposite sign. Figure 6.4 shows that the transients for charge and discharge, however, have very different time constants. This is because the interfacial area a takes on different values depending on the sign of the current input. To accurately capture this effect requires a time-varying model where the state-variable matrices take on different values based on the sign of the input. The linear model also predicts that the response simply scales with increasing charge/discharge current. In practice, however, Butler–Volmer nonlinearities change the response shape for high-rate charge and discharge. In [50], errors between the linear and nonlinear models are shown to emerge for charge/discharge rates greater than $0.2C$.

The distributions of acid concentration c, electrical potential in electrolyte ϕ_e, and electric potential in the solid phase ϕ_s after 900 s of $0.1C$ charge and discharge are shown in

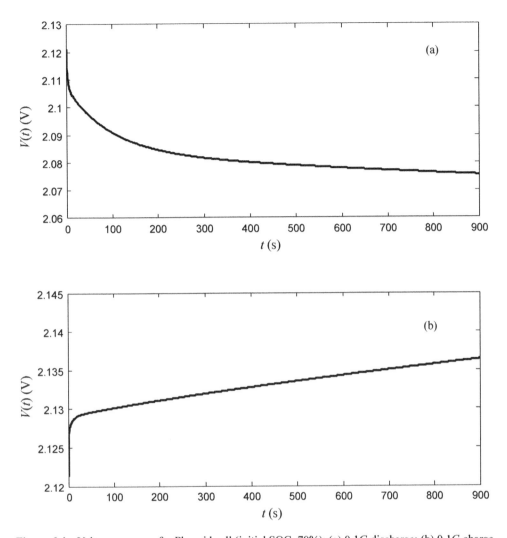

Figure 6.4 Voltage response for Pb–acid cell (initial SOC: 70%): (a) $0.1C$ discharge; (b) $0.1C$ charge

Figures 6.5 and 6.6, respectively. The concentration time responses for charge and discharge are shown in Figure 6.7. At $t = 0$, the acid concentration is uniform across the cell at 4.78 mol/L. For the discharge case, acid is consumed in both the positive and negative electrodes. Acid is not consumed in the separator, but diffuses from the separator into the electrodes, causing a hump in the middle of the concentration distribution. The acid is consumed more quickly in the positive electrode than in the negative electrode due to the larger a_2 and a in the positive electrode; so, the overall distribution slopes toward the positive electrode. If the discharge current is turned off at $t = 900$ s, then the acid concentration would eventually return to a flat distribution at the average concentration. The zero-flux boundary conditions are clearly enforced in the concentration response, giving zero slope at $x = 0$ and L. During

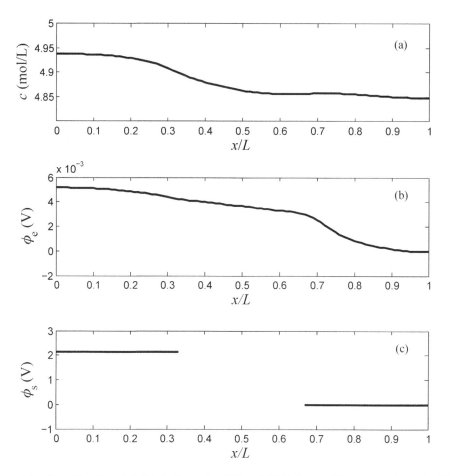

Figure 6.5 Pb–acid cell spatial distributions after 900 s of $0.1C$ charge: (a) concentration, $c(x, 900)$; (b) electrolyte potential, $\phi_e(x, 900)$; (c) electrode potential, $\phi_s(x, 900)$

charge, acid concentration increases in both the positive and negative electrodes, but again at a faster rate in the positive electrode. The higher concentration acid slowly diffuses into the separator, where there is no acid generation. Thus, the concentration distribution dips in the middle and generally slopes toward the negative electrode. The electrolyte and solid-phase potential do not change much during the simulation. The reference voltage for all of the potential calculations is the average electrolyte potential, which is constrained to be zero because the expansion in Equation (6.14) does not include a constant or bias term. The difference between $\phi_s(0, t)$ and $\phi_s(L, t)$ matches the 2.1 V observed in the voltage response. While $\phi_s(0, t)$ is always larger than $\phi_s(L, t)$ due to the potential developed at the solid–electrolyte interface, the $\phi_e(x, t)$ distribution slopes in the direction of ion transport, indicating the resistance-like effect of electrolyte diffusion. The zero-flux boundary conditions on electrolyte diffusion are clearly being enforced at $x = 0$ and L.

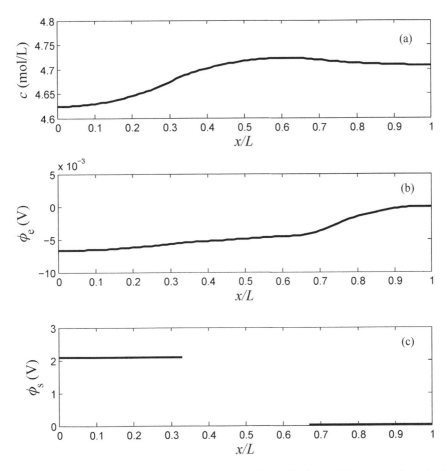

Figure 6.6 Pb–acid cell spatial distributions after 900 s of $0.1C$ discharge: (a) concentration, $c(x, 900)$; (b) electrolyte potential, $\phi_e(x, 900)$; (c) electrode potential, $\phi_s(x, 900)$

Figure 6.8 shows a switcher cycle response of the Ritz model. The current input is representative of the power demands of a locomotive moving railcars in a switchyard [50]. The current input consists of a 1.6 A (0.023 C) constant discharge associated with hotel loads (cooling, electronics, etc.) and short pulses of -16 A (0.229 C) and 12 A (0.171 C) corresponding to tractive effort and regenerative braking, respectively. This 7 min cycle is repeated until the pack is recharged at the end of a shift. The pack is slowly discharging due to the hotel loads and because the tractive effort energy is more than the regenerative braking during each cycle. Thus, the voltage (and SOC) are slowly decreasing, as seen in the simulation results. The voltage transients are primarily due to the double-layer effect. In this simulation, a constant specific area equal to the discharge value is used because the current is predominantly discharging during the switcher cycle. In this case it may not be necessary to include the complexity of time-varying properties in the Ritz model to get good results.

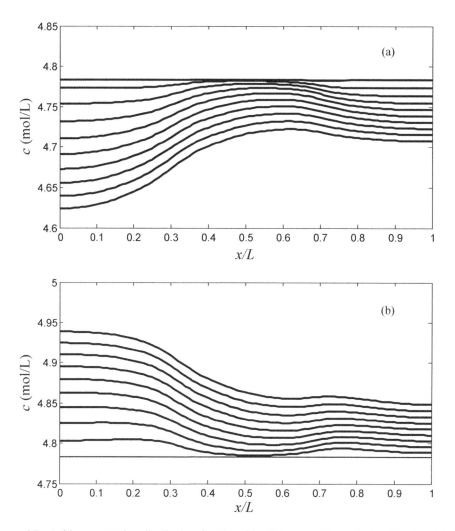

Figure 6.7 Acid concentration distribution for Pb–acid cell (every 100 s until $t = 900$ s): (a) $0.1C$ discharge; (b) $0.1C$ charge

6.2 Lithium-Ion Battery Model

Figure 6.9 shows a schematic diagram of the Li-ion cell model. The one-dimensional section from the negative current collector ($x = 0$) to the positive current collector ($x = L$) consists of three domains: the negative composite electrode (width δ_-), separator (width δ_{sep}), and positive composite electrode (width δ_+). During discharge, positively charged lithium ions diffuse to the surface of the Li_xC_6 active material particles that constitute the solid phase in the negative electrode. They undergo electrochemical reaction and transfer into a liquid or gelled electrolyte solution that makes up the electrolyte phase. The lithium ions travel through the electrolyte solution via diffusion and ionic conduction to the positive electrode, where

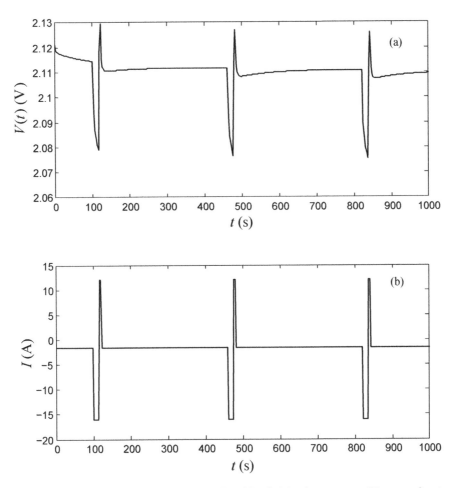

Figure 6.8 Switcher cycle simulation for Pb–acid cell: (a) voltage output; (b) current input

they react and diffuse towards the inner regions of metal oxide active material particles that constitute the solid phase of the positive electrode. The porous separator acts as an electronic insulator, forcing electrons to follow an opposite path through an external circuit or load. A review of Li-ion battery modeling is provided in [51] and details and background on the model presented in this section are in [25, 52, 53].

Four PDEs govern the dynamics of Li-ion batteries: conservation of species and charge in the electrode and electrolyte. They are linked by the Butler–Volmer equation. The model presented here is often termed a pseudo-two-dimensional model because one dimension is x and the other is the radial dimension in the spherical particles r. The particles are assumed to be distributed throughout the electrodes and modeled as a particles embedded in the electrode at each value of x. Thus, at each x there is also a radial coordinate r corresponding to the particle embedded at that point. It is called a *pseudo*-two-dimensional model because the neighboring particles are not directly coupled, unlike most two-dimensional PDEs (e.g., heat conduction in

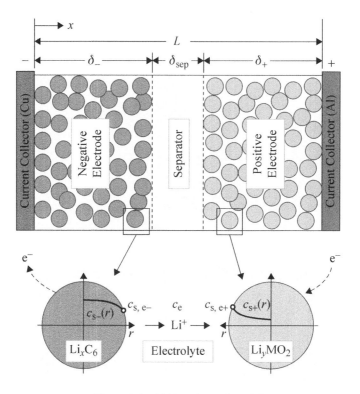

Figure 6.9 Lithium-ion cell model

a rectangular domain where the temperature $T(x, y)$ is coupled to neighboring temperatures in both the x and y directions). The embedded particles couple to the electrode through the r direction and the electrode is coupled through the x direction but there is not a direct path for ions to flow from inside one particle to an adjacent particle.

6.2.1 Conservation of Species

The composite electrodes are modeled using porous electrode theory, meaning that the solid-phase particles are assumed to be uniformly distributed throughout the electrode. Conservation of lithium in a single spherical active material particle is described by Fick's law of diffusion,

$$\frac{\partial c_s}{\partial t} = \frac{D_s}{r^2} \frac{\partial}{\partial r} \left(r^2 \frac{\partial c_s}{\partial r} \right) \quad \text{for } r \in (0, R_s), \tag{6.23}$$

where $r \in (0, R_s)$ is the radial coordinate, $c_s(r, t)$ is the concentration of lithium ions in the particle as a function of radial position and time, and D_s is the solid-phase diffusion coefficient.

We use the subscripts "s," "e," and "s,e" to indicate solid phase, electrolyte phase, and solid–electrolyte interface, respectively. The boundary conditions are

$$\left.\frac{\partial c_s}{\partial r}\right|_{r=0} = 0, \tag{6.24}$$

$$D_s \left.\frac{\partial c_s}{\partial r}\right|_{r=R_s} = -\frac{j}{F}, \tag{6.25}$$

where $j(x, t)$ is the rate of electrochemical reaction at the particle surface (with $j > 0$ indicating ion discharge) and F (96 487 C/mol) is the Faraday constant. Equations (6.23)–(6.25) are applied on a continuum basis across each electrode, giving solid-phase concentration a two-dimensional spatial dependency; that is, $c_s(x, r, t)$. The electrochemical model depends only upon concentration at the particle surface, $c_{s,e}(x, t) = c_s(x, R_s, t)$.

Conservation of lithium in the electrolyte phase yields

$$\varepsilon_e \frac{\partial c_e}{\partial t} = D_e^{\text{eff}} \frac{\partial^2 c_e}{\partial x^2} + \frac{a_s(1 - t_+^0)}{F} j \quad \text{for } x \in (0, L), \tag{6.26}$$

where $c_e(x, t)$ is the electrolyte phase lithium concentration, ε_e is the electrolyte phase volume fraction, t_+^0 is the transference number of Li^+ with respect to the velocity of solvent, and a_s the specific interfacial surface area. For the spherical active material particles occupying electrode volume fraction ε_s, the interfacial surface area is $a_s = 3\varepsilon_s/R_s$. The effective diffusion coefficient is calculated from a reference coefficient using the Bruggeman relation $D_e^{\text{eff}} = D_e \varepsilon_e^p$ that accounts for the tortuous path that Li^+ ions follow through the porous media. We assume that ε_e, t_+^0, and D_e^{eff} are constant within the negative electrode, separator, and positive electrode but can have different values in the three domains. Equation (6.26) has zero-flux boundary conditions at the current collectors:

$$\left.\frac{\partial c_e}{\partial x}\right|_{x=0} = \left.\frac{\partial c_e}{\partial x}\right|_{x=L} = 0. \tag{6.27}$$

6.2.2 Conservation of Charge

Charge conservation in the solid phase of each electrode is described by Ohm's law:

$$\sigma^{\text{eff}} \frac{\partial^2 \phi_s}{\partial x^2} - a_s j = 0 \quad \text{for } x \in (0, L), \tag{6.28}$$

where $\phi_s(x, t)$ and σ^{eff} are the potential and effective conductivities of the solid matrix, respectively, with σ^{eff} evaluated from active material reference conductivity σ as $\sigma^{\text{eff}} = \sigma \varepsilon_s$. Conductivity is assumed constant in the negative electrode ($\sigma^{\text{eff}} = \sigma_-^{\text{eff}}$ for $x \in (0, \delta_-)$), the separator, and the positive electrode ($\sigma^{\text{eff}} = \sigma_+^{\text{eff}}$ for $x \in (\delta_- + \delta_{\text{sep}}, L)$). The boundary conditions at the current collectors are proportional to applied current:

$$-\sigma_-^{\text{eff}} \frac{\partial \phi_s}{\partial x}\bigg|_{x=0} = \sigma_+^{\text{eff}} \frac{\partial \phi_s}{\partial x}\bigg|_{x=L} = \frac{I}{A}, \tag{6.29}$$

where A is electrode plate area and $I(t)$ is the applied current following the sign convention that a positive current discharges the battery. The boundary conditions at the separator require zero electronic current:

$$\frac{\partial \phi_s}{\partial x}\bigg|_{x=\delta_-} = \frac{\partial \phi_s}{\partial x}\bigg|_{x=\delta_- + \delta_{\text{sep}}} = 0. \tag{6.30}$$

Electrolyte phase charge conservation yields

$$\kappa^{\text{eff}} \frac{\partial^2 \phi_e}{\partial x^2} + \kappa_d^{\text{eff}} \frac{\partial^2 c_e}{\partial x^2} + a_s j = 0 \quad \text{for } x \in (0, L), \tag{6.31}$$

where $\phi_e(x, t)$ is the electrolyte phase potential and κ^{eff} the effective ionic conductivity, calculated from the Bruggeman relation $\kappa^{\text{eff}} = \kappa \varepsilon_e^p$.

The boundary conditions for Equation (6.31) are zero flux at the two current collectors:

$$\frac{\partial \phi_e}{\partial x}\bigg|_{x=0} = \frac{\partial \phi_e}{\partial x}\bigg|_{x=L} = 0. \tag{6.32}$$

At the electrode separator interfaces, we have flux continuity as

$$\left(\kappa^{\text{eff}} \frac{\partial \phi_e}{\partial x} + \kappa_d^{\text{eff}} \frac{\partial c_e}{\partial x}\right)\bigg|_{x_-} = \left(\kappa^{\text{eff}} \frac{\partial \phi_e}{\partial x} + \kappa_d^{\text{eff}} \frac{\partial}{\partial x} c_e\right)\bigg|_{x_+}, \tag{6.33}$$

where $x = \delta_-$ and $\delta_- + \delta_{\text{sep}}$ corresponding to the two separator interfaces.

6.2.3 Reaction Kinetics

The four governing PDEs (6.23), (6.26), (6.28), and (6.31) describing field variables, $c_{s,e}(x, t)$, $c_e(x, t)$, $\phi_s(x, t)$, and $\phi_e(x, t)$ are coupled by the Butler–Volmer electrochemical kinetic expression

$$j = i_0 \left[\exp\left(\frac{\alpha_a F}{RT}\eta\right) - \exp\left(-\frac{\alpha_c F}{RT}\eta\right) \right] \quad \text{for } x \in (0, L), \tag{6.34}$$

where $i_0(x, t)$ is the exchange current density, $\eta(x, t)$ is the overpotential, and α_a and α_c are the anodic and cathodic transfer coefficients, respectively. The exchange current density is related to both solid surface and electrolyte concentrations according to

$$i_0 = k(c_e)^{\alpha_a}(c_{s,\text{max}} - c_{s,e})^{\alpha_a}(c_{s,e})^{\alpha_c} \quad \text{for } x \in (0, L), \tag{6.35}$$

where k is a kinetic rate constant and $c_{s,max}$ is the maximum solid-phase lithium concentration. In Equation (6.34), j is driven by overpotential, defined as the difference between solid and electrolyte phase potentials minus the thermodynamic equilibrium potential U of the solid phase:

$$\eta = \phi_s - \phi_e - U \quad \text{for } x \in (0, L). \tag{6.36}$$

Equilibrium potential $U(c_{s,e})$ is evaluated as a function of the solid phase concentration at the particle surface and has different values in the two electrodes.

6.2.4 Cell Voltage

With boundary conditions applied galvanostatically as in Equation (6.29), cell current $I(t)$ is the model input. Voltage across the cell terminals is calculated from

$$V(t) = \phi_s(L, t) - \phi_s(0, t) - \frac{R_f}{A} I(t), \tag{6.37}$$

where R_f is an empirical contact resistance.

6.2.5 Linearization

The Butler–Volmer equation (6.34) must be linearized at an equilibrium point in order to produce a linear model. As a first step in the linearization process, the equilibrium distributions for concentrations and potentials are calculated. At equilibrium, the currents $j = I = 0$, so $\eta = 0$ from the Butler–Volmer equation and constant distributions (independent of space and time) satisfy the governing equations. Specifically, $c_s(r, t) = \bar{c}_s = $ constant (independent of r and t) satisfies conservation of lithium in the solid phase (6.23) and boundary conditions (6.24) and (6.25) because $\bar{c}_s = 0$. Thus, the equilibrium lithium concentration is uniform throughout the particle and surface concentration equals the average concentration, so $c_{s,e}(x, t) = c_s(r, t) = \bar{c}_s$. Similarly, conservation of lithium in the electrolyte phase (6.26) and boundary conditions (6.27) are satisfied with $c_e(x, t) = \bar{c}_e = $ constant (independent of x and t). It is safe to assume that $\bar{c}_e = 0$ because in equilibrium the lithium ions are stored in either the positive or negative electrode with very few remaining in the electrolyte. Charge concentration in the solid phase at equilibrium is also constant with $\phi_s(x, t) = \bar{\phi}_s$. A constant electrolyte potential $\phi_e(x, t) = \bar{\phi}_e$ satisfies charge concentration in the electrolyte phase in equilibrium.

From the definition of overpotential in Equation (6.36), we have the equilibrium relationships

$$\bar{\phi}_s^- = \bar{U}^- + \bar{\phi}_e, \tag{6.38}$$

$$\bar{\phi}_s^+ = \bar{U}^+ + \bar{\phi}_e, \tag{6.39}$$

where $\bar{U} = U(\bar{c}_s)$. If we assign the negative terminal as ground, then $\bar{\phi}_s^- = 0$, $\bar{\phi}_e = -\bar{U}^-$, and

$$\bar{V} = \bar{\phi}_s^+ = \bar{U}^+ - \bar{U}^- \tag{6.40}$$

equals the open-circuit voltage.

In summary, the equilibrium variables for an Li-ion cell are determined by the specified or given value of the average concentration $\bar{c}_s$. As we will see in Chapter 7, knowing $\bar{c}_s$ is equivalent to knowing the SOC of the cell. Given the SOC or $\bar{c}_s$, the equilibrium values of V, ϕ_s, and ϕ_e can all be calculated.

The second step in linearization is to use perturbation equations that set each variable equal to its equilibrium value plus a small deviation, indicated by the variable with a tilde on top (e.g., $c_s(x, t) = \bar{c}_s + \tilde{c}_s(x, t)$, where $\tilde{c}_s(x, t)$ is small). For the variables with zero equilibrium values ($\eta(x, t)$, $c_e(x, t)$, $j(x, t)$, and ϕ_s^-) the tilde variables equal the original values (e.g., $\eta(x, t) = \tilde{\eta}(x, t)$), so we leave off the tildes for simplicity.

Substitution of the perturbation equations into the governing equations, expanding nonlinear terms using a Taylor series, canceling the equilibrium terms, and keeping only first-order terms in the tilde variables results in a set of linear equations. For the Li-ion model, all of the equations are linear with the exception of the Butler–Volmer equation (6.34) and the overpotential equation (6.36). For the linear equations, one can simply substitute all variables with tilde variables to obtain the "linearized" equations. The nonlinear Equation (6.34) linearizes to

$$\eta = R_{ct} j, \qquad (6.41)$$

with charge-transfer resistance $R_{ct} = RT/[\bar{i}_0 F(\alpha_a + \alpha_c)]$ and $\bar{i}_0$ is calculated at $c_e = 0$ and $c_{s,e} = \bar{c}_s$. Equation (6.36) linearizes to

$$\eta = \tilde{\phi}_s - \tilde{\phi}_e - \tilde{U}, \qquad (6.42)$$

where

$$\tilde{U} = \frac{\partial U}{\partial c} \tilde{c}_{s,e} \qquad (6.43)$$

with

$$\frac{\partial U}{\partial c} = \frac{\partial U}{\partial c_{s,e}} \Big|_{c_{s,e} = \bar{c}_s} \qquad (6.44)$$

assumed constant.

6.2.6 Impedance Solution

For the pseudo-two-dimensional model of an Li-ion cell presented here, the embedded particles complicate the problem to the point where an analytical solution is not possible. In this section, we neglect electrolyte diffusion in order to obtain an analytical solution for the remaining variables, including the current density distribution $j(x, t)$. Using this distribution as input to an FEM model of electrolyte diffusion allows an approximate inclusion of this important effect. The approach presented in this section follows that of [25].

The linearized particle diffusion equation is

$$\frac{\partial \tilde{c}_s}{\partial t} = \frac{D_s}{r^2} \frac{\partial}{\partial r} \left(r^2 \frac{\partial \tilde{c}_s}{\partial r} \right) \tag{6.45}$$

with the boundary conditions

$$\left. \frac{\partial \tilde{c}_s}{\partial r} \right|_{r=0} = 0 \tag{6.46}$$

and

$$D_s \left. \frac{\partial \tilde{c}_s}{\partial r} \right|_{r=R_s} = -\frac{j}{F}. \tag{6.47}$$

Taking the Laplace transform of Equation (6.45) subject to the boundary conditions yields the transfer function

$$\frac{\tilde{C}_{s,e}(x, s)}{J(x, s)} = \frac{1}{F} \left(\frac{R_s}{D_s} \right) \left[\frac{\tanh(\beta)}{\tanh(\beta) - \beta} \right] = \mathcal{G}_p(s), \tag{6.48}$$

where $\tilde{C}_{s,e}(x, s)$ and $J(x, s)$ are the Laplace transforms of $\tilde{c}_{s,e}(x, t)$ and $j(x, t)$, respectively, and $\beta = R_s \sqrt{s/D_s}$. This is the spherical particle transfer function that we derived in Chapter 4 and simulated in Chapter 5.

The Laplace transform of the linearized Butler–Volmer equation (6.41) is

$$\mathcal{N}(x, s) = R_{ct} J(x, s), \tag{6.49}$$

where $\mathcal{N}(x, s) = \mathcal{L}\{\eta(x, t)\}$.

If we neglect electrolyte diffusion, then the remaining variables of interest are ($c_{s,e}$, ϕ_e, and ϕ_s). Under this assumption, the positive and negative electrodes are decoupled from one another. The separator does not contribute to the analytical solution because there are no particles or electrodes. We therefore seek analytical solutions for $c_{s,e}$, ϕ_e, and ϕ_s in a single electrode and define the dimensionless spatial variable $z = x/\delta$, where δ is the electrode thickness and $z = 0$ and 1 at the current collector and separator interfaces, respectively.

The Laplace transform of the solid-phase charge conservation equation (6.28) is

$$\frac{\sigma^{\text{eff}}}{\delta^2} \frac{\partial^2 \tilde{\Phi}_s(z, s)}{\partial z^2} - J(z, s) = 0 \tag{6.50}$$

with $x \rightarrow z$. The boundary conditions are

$$-\frac{\sigma^{\text{eff}}}{\delta} \left. \frac{\partial \tilde{\Phi}_s}{\partial z} \right|_{z=0} = \frac{\mathcal{I}}{A}, \tag{6.51}$$

where $\mathcal{I}(s) = \mathcal{L}\{I(t)\}$ and

$$\left.\frac{\partial \tilde{\Phi}_{\mathrm{s}}}{\partial z}\right|_{z=1} = 0. \tag{6.52}$$

Neglecting electrolyte diffusion, the Laplace transform of the electrolyte charge conservation equation (6.31) becomes

$$\frac{\kappa^{\mathrm{eff}}}{\delta^2} \frac{\partial^2 \tilde{\Phi}_{\mathrm{e}}}{\partial z^2} + J = 0 \tag{6.53}$$

with the boundary condition at the current collector

$$\left.\frac{\partial \tilde{\Phi}_{\mathrm{e}}}{\partial z}\right|_{z=0} = 0. \tag{6.54}$$

The boundary condition at the separator can be obtained by integration of the charge conservation equation over the domain, or, equivalently, enforcing charge conservation in the electrode as a whole. Integration of the solid-phase charge conservation equation (6.50) gives

$$\int_0^1 J \, \mathrm{d}z = \int_0^1 \frac{\sigma^{\mathrm{eff}}}{\delta^2} \frac{\partial^2 \tilde{\Phi}_{\mathrm{s}}}{\partial z^2} \, \mathrm{d}z = \left.\frac{\sigma^{\mathrm{eff}}}{\delta^2} \frac{\partial \tilde{\Phi}_{\mathrm{s}}}{\partial z}\right|_0^1 = -\frac{\mathcal{I}}{A\delta}, \tag{6.55}$$

using the boundary conditions. From liquid-phase charge conservation, (6.53),

$$\int_0^1 J \, \mathrm{d}z = -\int_0^1 \frac{\kappa^{\mathrm{eff}}}{\delta^2} \frac{\partial^2 \tilde{\Phi}_{\mathrm{e}}}{\partial z^2} \, \mathrm{d}z = -\frac{\kappa^{\mathrm{eff}}}{\delta^2} \frac{\partial \tilde{\Phi}_{\mathrm{e}}}{\partial z}(1, s), \tag{6.56}$$

using the zero flux boundary condition at $z = 0$. Equating Equations (6.55) and (6.56) provides the missing boundary condition on electrolyte phase potential at the separator:

$$\frac{\kappa^{\mathrm{eff}}}{\delta} \frac{\partial \tilde{\Phi}_{\mathrm{e}}}{\partial z}(1, s) = \frac{\mathcal{I}}{A}. \tag{6.57}$$

The last equation needed for the analytical solution is the Laplace transform of Equation (6.42):

$$\mathcal{N} = \tilde{\Phi}_{\mathrm{s}} - \tilde{\Phi}_{\mathrm{e}} - \frac{\partial U}{\partial c} \tilde{C}_{\mathrm{s,e}}, \tag{6.58}$$

which depends only on the difference between the solid and electrolyte phase potentials $\tilde{\Phi}_{\mathrm{s-e}} = \Phi_{\mathrm{s}} - \Phi_{\mathrm{e}}$. Combining Equations (6.50) and (6.53), we obtain

$$\frac{\partial^2 \tilde{\Phi}_{\mathrm{s-e}}}{\partial z^2} = \delta^2 \left(\frac{1}{\kappa^{\mathrm{eff}}} + \frac{1}{\sigma^{\mathrm{eff}}}\right) J \tag{6.59}$$

with boundary conditions

$$-\frac{\sigma^{\text{eff}}}{\delta}\frac{\partial\tilde{\Phi}_{\text{s-e}}}{\partial z}(0,s) = \frac{\kappa^{\text{eff}}}{\delta}\frac{\partial\tilde{\Phi}_{\text{s-e}}}{\partial z}(1,s) = \frac{I}{A}, \tag{6.60}$$

obtained by combining the solid and electrolyte phase potential boundary conditions.

Equation (6.58) can be simplified using the transfer function (6.48) and linearized Butler–Volmer equation (6.49) to produce

$$\tilde{\Phi}_{\text{s-e}} = \left[R_{\text{ct}} + \frac{\partial U}{\partial c}\mathcal{G}_p\right]\tilde{J}. \tag{6.61}$$

Combining Equations (6.59) and (6.61), we obtain a single ODE:

$$\frac{\partial^2\tilde{\Phi}_{\text{s-e}}}{\partial z^2} - \delta^2\left(\frac{1}{\kappa^{\text{eff}}} + \frac{1}{\sigma^{\text{eff}}}\right)\left[R_{\text{ct}} + \frac{\partial U}{\partial c}\mathcal{G}_p(s)\right]^{-1}\tilde{\Phi}_{\text{s-e}} = 0, \tag{6.62}$$

with boundary conditions (6.60) in the single unknown $\tilde{\Phi}_{\text{s-e}}(x,s)$. The beauty of the transfer function approach taken here is that in the ODE (6.62) the Laplace variable s is a parameter, so one need only solve the linear, constant-parameter equation

$$\frac{\partial^2\tilde{\Phi}_{\text{s-e}}}{\partial z^2} - v^2\tilde{\Phi}_{\text{s-e}} = 0, \tag{6.63}$$

where

$$v(s) = \delta\left(\frac{1}{\kappa^{\text{eff}}} + \frac{1}{\sigma^{\text{eff}}}\right)^{1/2}\left[R_{\text{ct}} + \frac{\partial U}{\partial c}\mathcal{G}_p(s)\right]^{-1/2} \tag{6.64}$$

is independent of $\tilde{\Phi}_{\text{s-e}}$ (linear) and z (constant parameter).

The solutions of Equation (6.63) are exponentials of the form

$$\tilde{\Phi}_{\text{s-e}}(z,s) = C_1(s)\sinh[v(s)z] + C_2(s)\cosh[v(s)z]. \tag{6.65}$$

Substitution of Equation (6.65) into the boundary conditions (6.60) yields the coefficients

$$\frac{C_1(s)}{I(s)} = -\frac{\delta}{v(s)A\sigma^{\text{eff}}}, \tag{6.66}$$

$$\frac{C_2(s)}{I(s)} = \frac{\delta[\kappa^{\text{eff}}\cosh(v(s)) + \sigma^{\text{eff}}]}{A\kappa^{\text{eff}}\sigma^{\text{eff}}v(s)\sinh(v(s))}. \tag{6.67}$$

Substitution of the coefficients (6.66) into Equation (6.65) yields

$$\frac{\tilde{\Phi}_{\text{s-e}}(z,s)}{I(s)} = \frac{\delta}{Av\sinh v}\left\{\frac{1}{\sigma^{\text{eff}}}\cosh[v(z-1)] + \frac{1}{\kappa^{\text{eff}}}\cosh[vz]\right\}. \tag{6.68}$$

Using Equation (6.61), we obtain the transfer function

$$\frac{J(z,s)}{\mathcal{I}(s)} = \frac{J(z,s)}{\tilde{\Phi}_{s-e}(z,s)} \frac{\tilde{\Phi}_{s-e}(z,s)}{\mathcal{I}(s)} = \frac{v^2 \sigma^{\text{eff}} \kappa^{\text{eff}}}{\delta^2(\sigma^{\text{eff}} + \kappa^{\text{eff}})} \frac{\tilde{\Phi}_{s-e}(z,s)}{\mathcal{I}(s)}$$

$$= \frac{v}{\delta A(\kappa^{\text{eff}} + \sigma^{\text{eff}}) \sinh v} \{\kappa^{\text{eff}} \cosh[v(z-1)] + \sigma^{\text{eff}} \cosh[v(z)]\}. \quad (6.69)$$

From Equation (6.49) we have

$$\frac{\mathcal{N}(z,s)}{\mathcal{I}(s)} = R_{\text{ct}} \frac{J(z,s)}{\mathcal{I}(s)} \quad (6.70)$$

and using Equation (6.48)

$$\frac{\tilde{C}_{s,e}(z,s)}{\mathcal{I}(s)} = \frac{\tilde{C}_{s,e}(s)}{J(s)} \frac{J(z,s)}{\mathcal{I}(s)}, \quad (6.71)$$

both of which use the transfer function (6.69).

6.2.7 FEM Electrolyte Diffusion

Now we reintroduce electrolyte diffusion using an FEM model that allows relaxation of the simplifying assumption used to obtain an analytical solution in the previous section. The current density solution in Equation (6.69) is the input to an FEM electrolyte diffusion model. Electrolyte diffusion correction terms are calculated using the FEM model that adds the effects of electrolyte diffusion to electrolyte potential, and hence voltage.

Equation (6.26) governing conservation of lithium in the electrolyte was not used in the impedance model and the concentration coupling term in the electrolyte equation (6.31) was neglected. Using the FEM method described in Chapter 4, we discretize these two equations to

$$\mathbf{M}\dot{\mathbf{c}}_e = -\mathbf{K}\mathbf{c}_e + \mathbf{F}\mathbf{j} \quad (6.72)$$

and

$$\mathbf{K}_\phi \boldsymbol{\phi}_e + \mathbf{K}_c \mathbf{c}_e(t) = \mathbf{F}_\phi \mathbf{j}, \quad (6.73)$$

where $\mathbf{c}_e(t)$ and $\boldsymbol{\phi}_e(x,t)$ are the nodal electrolyte concentrations $\tilde{c}_e(x_i,t)$ and potentials $\tilde{\phi}_e(x_i,t)$ and

$$\mathbf{j}^{\text{T}}(t) = [j_-(x_1,t), \ldots j_-(x_{n_-},t), 0, \ldots 0, \; j_+(x_{n_{\text{cell}}-n_++1},t), \ldots j_+(x_{n_{\text{cell}}},t)] \quad (6.74)$$

is the current density calculated at the n_- nodal points in the negative electrode and n_+ nodal points in the positive electrode using the transfer function (6.69). The current density is zero for the nodal points in the separator.

The electrolyte concentration distribution is calculated by taking the Laplace transform of Equation (6.72) and solving for $\mathbf{C}_e(s) = \mathcal{L}(\mathbf{c}_e(t))$ as

$$\frac{\mathbf{C}_e(s)}{\mathcal{I}(s)} = (\mathbf{K} + s\mathbf{M})^{-1}\mathbf{FJ}, \tag{6.75}$$

where $\mathbf{J}_i = J(z_i, s)/\mathcal{I}(s)$.

Solution of the discretized electrolyte potential equation (6.73) requires inversion of the matrix $\mathbf{K}_\phi$. This matrix is singular, however, owing to the zero-flux boundary conditions at $x = 0$ and L. To avoid this problem, we define voltages relative to $\phi_e(0, t)$ so that $\Delta\phi_e(x, t) = \phi_e(x, t) - \phi_e(0, t)$ is calculated. Relative potential is all that is required to calculate the voltage. It is not possible to enforce both $\Delta\phi_e(0, t) = 0$ and $\Delta\phi_e'(0, t) = 0$ for a second-order ODE, however, so we approximate this by subtracting the $(1, 1)$ element of $\mathbf{K}_\phi$ from the first column of all rows of $\mathbf{K}_\phi$ as follows:

$$\mathbf{K}_{\Delta\phi} = \mathbf{K}_\phi - (\mathbf{K}_\phi)_{1,1}\begin{bmatrix} 1 & 0 & \cdots & 0 \\ 1 & 0 & & \\ \vdots & & \ddots & \\ 1 & 0 & \cdots & 0 \end{bmatrix}, \tag{6.76}$$

to produce an approximation of

$$\frac{\Delta\Phi_e(s)}{\mathcal{I}(s)} = (\mathbf{K}_{\Delta\phi})^{-1}\left(-\mathbf{K}\frac{\Delta\mathbf{C}_e(s)}{\mathcal{I}(s)} + \mathbf{F}\frac{\Delta\mathbf{J}}{\mathcal{I}(s)}\right), \tag{6.77}$$

where $\Delta\mathbf{C}_e = \mathbf{C}_e - (\mathbf{C}_e)_{1,1}$ and $\Delta\mathbf{J} = \mathbf{J} - (\mathbf{J})_{1,1}$.

6.2.8 Overall System Transfer Function

The voltage equation (6.37) can be expanded as

$$\tilde{V}(t) = \tilde{\phi}_e(L, t) - \tilde{\phi}_e(0, t) + \eta(L, t) - \eta(0, t)$$
$$+ \frac{\partial U^+}{\partial c}\tilde{c}_{s,e}(L, t) - \frac{\partial U^-}{\partial c}\tilde{c}_{s,e}(0, t) - \frac{R_f}{A}I(t). \tag{6.78}$$

After application of the Laplace transform, the final, overall system impedance is

$$\frac{\tilde{V}(s)}{\mathcal{I}(s)} = \frac{\Delta\Phi_e(L, s)}{\mathcal{I}(s)} + \frac{\Delta\mathcal{N}(L, s)}{\mathcal{I}(s)} + \frac{\partial U^+}{\partial c}\frac{\tilde{C}_{s,e}(L, s)}{\mathcal{I}(s)} - \frac{\partial U^-}{\partial c}\frac{\tilde{C}_{s,e}(0, s)}{\mathcal{I}(s)} - \frac{R_f}{A}, \tag{6.79}$$

where $\Delta\Phi_e(L, s)$ is the n_{cell}th element of $\Delta\Phi_e(s)$ and $\Delta\mathcal{N}(x, s) = \mathcal{N}(x, s) - \mathcal{N}(0, s)$.

6.2.9   Time-Domain Model and Simulation Results

The method described in the previous sections generates a transfer function model of an Li-ion cell. To be useful for time-domain simulation and analysis, the transcendental transfer functions must be discretized to polynomial transfer functions or state-space form. The system identification method described in Chapter 4 is used to generate a time-domain model from the frequency-domain data. The model parameters are given in Table 6.4 corresponding to an Li-ion cell in the literature [25] at an initial SOC of 50%.

Figure 6.10 shows the impedance frequency response for the transcendental transfer function and the system-identified time-domain model with $N = 8$. The magnitude and phase curves lie on top of one another up to the desired bandwidth of 10 Hz. In Figure 6.11 the discretized time-domain model is used to simulate the voltage response to a pulse charge/discharge current input. The voltage rises and falls relative to the open circuit $\bar{V} = 3.6$ V at 50% SOC.

Transcendental transfer functions for all of the variables of interest in the cell have been derived. The frequency response of these variables can be calculated and discretized. In many cases it is of interest to plot the time response of a distribution as snapshots of the variable plotted versus x at several different times t_i. This requires the development of transfer

Table 6.4   Parameters for Li-ion cell model

Parameter	Negative	Separator	Positive
Design specifications			
Thickness, δ (cm)	50×10^{-4}	25.4×10^{-4}	36.4×10^{-4}
Particle radius, R_s (cm)	1×10^{-4}		1×10^{-4}
Active material volume fraction, ε_s	0.580		0.500
Polymer phase volume fraction, ε_p	0.048	0.5	0.110
Conductive filler volume fraction, ε_f	0.040		0.06
Porosity, ε_e	0.332	0.5	0.330
Electrode plate area, A (cm^2)		10 452	
Current collector contact resistance, R_f (Ω cm^2)		20	
Li ion concentrations			
Maximum solid-phase concentration $c_{s,max}$ (mol/cm^3)	16.1×10^{-3}		23.9×10^{-3}
Stoichiometry at 0% SOC, $x_{0\%}$	0.126		0.936
Stoichiometry at 100% SOC, $x_{100\%}$	0.676		0.442
Average electrolyte concentration, $\bar{c}_e$ (mol/cm^3)		1.2×10^{-3}	
Kinetic and transport properties			
Exchange current density, i_0 (A/cm^2)	3.6×10^{-3}		2.6×10^{-3}
Charge-transfer coefficients, α_a, α_c	0.5, 0.5		0.5, 0.5
Solid-phase Li diffusion coefficient, D_s (cm^2/s)	2.0×10^{-12}		3.7×10^{-12}
Solid-phase conductivity, σ (S/cm)	1.0		0.1
Bruggeman porosity exponent, p	1.5	1.5	1.5
Electrolyte phase Li$^+$ diffusion coefficient, D_e (cm^2/s)		2.6×10^{-6}	
Electrolyte phase ionic conductivity, κ (S/cm)		$\kappa = 15.8c_e \exp[0.85(1000c_e)^{1.4}]$	
Electrolyte activity coefficient, $f_\pm$		1.0	
Li$^+$ transference number, t_+^0		0.363	

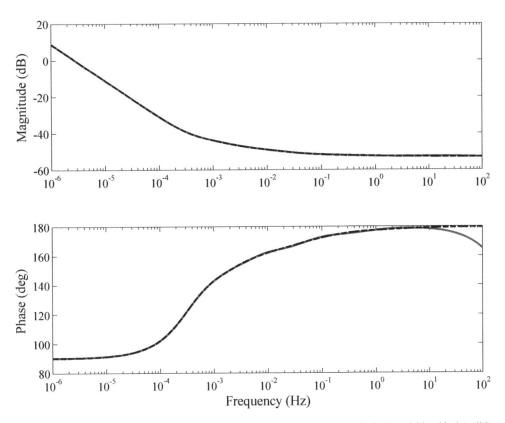

Figure 6.10 Li-ion cell impedance frequency response: transcendental (dashed) and identified (solid) transfer functions

functions at multiple values of x that can be simulated in time. This was done for 25 points evenly distributed along the x-axis for the $j(x, t)$, $c_{s,e}(x, t)$, and $c_e(x, t)$ transfer functions and the time responses are plotted in Figures 6.12, 6.13, and 6.14, respectively, for a $5C$ (30 A) discharge from 50% SOC initial condition. For 25 10th-order approximations, the simulation model order for each plot is 250 states. Model-order reduction could easily reduce the number of states because the dynamic characteristics of the transfer functions are very similar.

As shown in Figure 6.12, initial spikes in reaction current $j(x, t)$ near the separator decay as lithium is deinserted/inserted from the negative/positive electrode surface. Equilibrium potentials rise/fall most rapidly near the separator, penalizing further reaction, and over time $j(x, t)$ becomes more uniform. Surface concentrations $c_{s,e}(x, t)$, shown in Figure 6.13, fall/rise in a distributed manner consistent with the time history of reaction, $j(x, t)$. While discharge continues, $c_{s,e}(x, t)$ continues to rise/fall and never reaches steady state due to an electrode bulk concentration free integrator term. Figure 6.14 shows that electrolyte concentration $c_e(x, t)$ does approach a steady-state distribution due to offsetting source/sink terms in the $j(x, t)$ distribution in the negative/positive electrode regions.

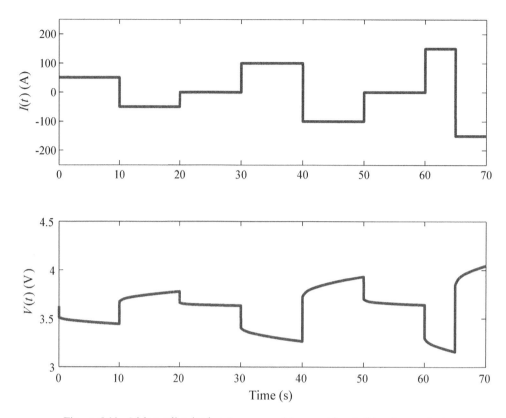

Figure 6.11   Li-ion cell pulse input response: (a) current input; (b) voltage output

6.3   Nickel-Metal Hydride Battery Model

Ni–MH battery cells have much in common with Li-ion cells. Both have a unary electrolyte
with one charge-carrying ion: Li^+ for Li-ion and OH^- for Ni–MH. As shown in Figure 6.15,
the cells are divided into three porous regions: negative electrode, separator, and positive
electrode. Active material particles are distributed throughout the electrodes. Unlike Li-ion
cells, however, the charge-carrying ion in the solid phase (H^+) is not the same as the charge-
carrying ion in the liquid phase (OH^-). Also, the particles in the positive electrode of an
Ni–MH cell are elongated, so researchers often approximate them as cylinders rather than the
spheres used in Li-ion cells (and the negative electrode of Ni–MH cells). Readers can refer to
[48, 54–57] for further information on Ni–MH modeling.

The Pb–acid and Li-ion models developed previously used Ritz and a combination of
analytical, FEM, and frequency-response fitting to discretize the governing PDEs. These
techniques have the advantages of incorporating all of the dynamic processes in the model,
converging rapidly with increasing model order, and being numerically efficient. They are
based, however, on linearization and can be difficult to apply, so a simpler approach that retains
key nonlinearities may be needed in some cases. For the Ni–MH model derived in this section,
we take a simplified approach called the single particle (SP) model [23, 58, 59]. In the SP

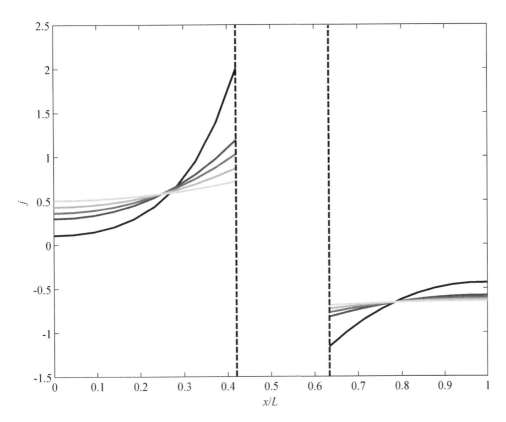

Figure 6.12 Li-ion current density distribution $j(x, t)$ time response: $5C$ discharge from 50% SOC at various times

model, a coarse discretization of the cell domain into three nodes (negative electrode, separator, and positive electrode) is used by assuming uniform distributions in these three regions, as shown in Figure 6.16. In addition, electrolyte diffusion is assumed to be instantaneous, so the electrolyte concentration $c_e(x, t)$ is uniform throughout the cell and $\partial c_e/\partial x = 0$. Conservation of species requires that while OH^- ions are inserted in one electrode and deinserted in the other, the average concentration $(c_e(x, t)$ in the SP model) is constant in time so $\partial c_e/\partial t = 0$ as well. These assumptions only hold, however, for small current charge/discharge or for an electrolyte with high ionic conductivity. Additional assumptions in the model include:

- Each electrode is a two-phase system consisting of the solid matrix and liquid electrolyte. Side reactions are neglected.
- The nickel electrode is modeled as cylindrical needles without a substrate inside. The porosity variation is neglected.
- The metal hydride electrode is modeled to consist of spherical particles with constant porosity.
- Thermal effects are discounted.

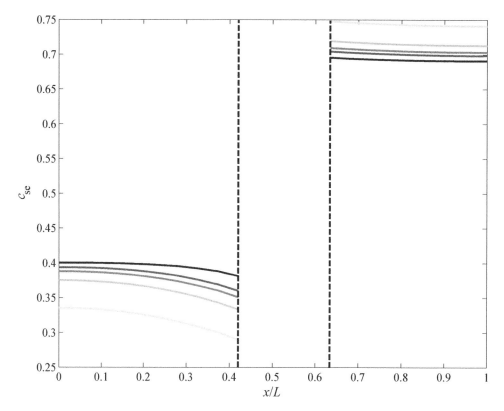

Figure 6.13 Li-ion solid-phase surface concentration distribution $c_{s,e}(x, t)$ time response: $5C$ discharge from 50% SOC at various times

6.3.1   Solid-Phase Diffusion

One species, either a proton or atomic hydrogen, is present in the solid phase of the nickel and metal hydride electrodes, respectively. This species is represented by a uniform symbol, H, in this model. The SP model has only two particles – one each in the positive and negative electrode. The governing equations of the spherical material particles in the negative, metal hydride electrode are the same as in the Li-ion model. Conservation of H in the cylindrical active material particle of the nickel, positive electrode is described by Fick's law of diffusion:

$$\frac{\partial c_s^+}{\partial t} = \frac{D_s^+}{r} \frac{\partial}{\partial r} \left(r \frac{\partial c_s^+}{\partial r} \right) \quad \text{for } r \in (0, R_s^+), \tag{6.80}$$

where $r \in (0, R_s^+)$ is the radial coordinate, $c_s^+(r, t)$ is the concentration of H$^+$ ions in the particle as a function of radial position and time, and D_s^+ is the solid-phase diffusion coefficient. We use the subscripts "s," "e," and "s,e" to indicate solid phase, electrolyte phase, and solid–electrolyte

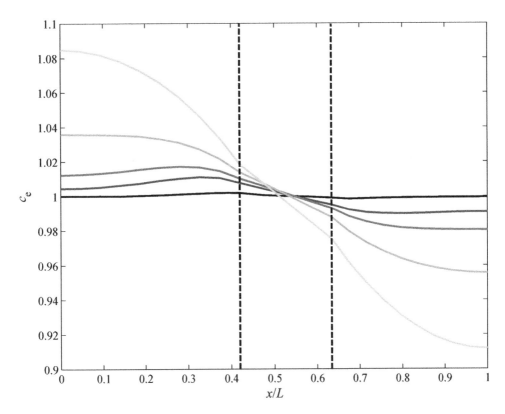

Figure 6.14   Li-ion electrolyte concentration distribution $c_e x, t$ time response: $5C$ discharge from 50% SOC at various times

interface, respectively. The boundary conditions are

$$\frac{\partial c_s^+}{\partial r}\bigg|_{r=0} = 0, \tag{6.81}$$

$$D_s^+ \frac{\partial c_s^+}{\partial r}\bigg|_{r=R_s^+} = \frac{j^+}{F}, \tag{6.82}$$

where $j^+(x, t)$ is the rate of electrochemical reaction at the particle surface, and F (96 487 C/mol) is the Faraday constant. The concentration at the particle surface $c_{s,e}^+(x, t) = c_s(R_s, t)$. The nickel electrode actually consists of composite cylindrical needles with a substrate inside. For simplicity, we approximate this as a solid cylinder. The cylinder radius is calculated to provide the same active material as the composite needle.

In Chapter 4 we derived the analytical spherical and cylindrical particle transfer functions and discretized them using a variety of methods to produce the polynomial transfer functions

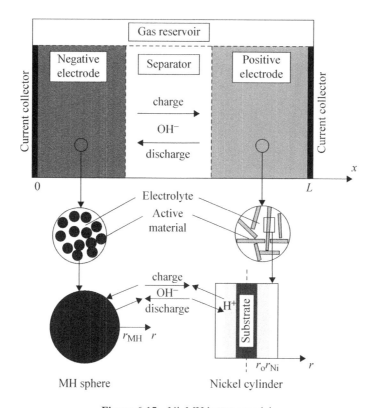

Figure 6.15   Ni–MH battery model

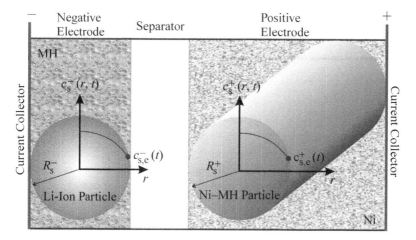

Figure 6.16   Ni–MH SP model

as follows:

$$\frac{C_{s,e}^-(s)}{J^-(s)} = G_{s,e}^-(s), \tag{6.83}$$

$$\frac{C_{s,e}^+(s)}{J^+(s)} = G_{s,e}^+(s), \tag{6.84}$$

where the $\pm$ sign indicates positive (cylindrical particle) and negative (spherical particle) electrodes. Using the IMA solution in Equation (4.108), for example,

$$G_{s,e}^-(s) = \frac{[R_s^-]^2 s + 9D_s^-}{3R_s^- D_s^- F s}, \tag{6.85}$$

$$G_{s,e}^+(s) = \frac{[R_s^+]^2 s + 6D_s^+}{3R_s^+ D_s^+ F s}. \tag{6.86}$$

Of course, a first-order transfer function will perform poorly in practice, so higher order approximations will be required for most cases.

6.3.2 Conservation of Charge

Charge conservation in the solid phase of the negative electrode is

$$j^-(t) = \frac{I(t)}{a_s^- A L^-}, \tag{6.87}$$

where L^- is the thickness of the negative electrode, A is electrode plate area, a_s^- is the specific interfacial area, and $I(t)$ is the applied current following the sign convention that a positive current discharges the battery. For the positive electrode:

$$j^+(t) = -\frac{I(t)}{a_s^+ A L^+}. \tag{6.88}$$

6.3.3 Reaction Kinetics

With constant electrolyte concentration, the Butler–Volmer kinetics for the negative electrode simplify to

$$j^-(t) = i_0^- \left\{ \frac{c_{s,e}^-(t)}{c_{s,ref}} \exp\left[\frac{\alpha^- F}{RT} \eta^-(t)\right] - \exp\left[-\frac{\alpha^- F}{RT} \eta^-(t)\right] \right\} \tag{6.89}$$

where i_0 is the exchange current density, $\eta^-(t)$ is the overpotential, α^+ is the transfer coefficient, and $c_{s,ref}$ is the reference H concentration. If we assume that $c_{s,e}^-(t) \approx c_{s,ref}$, then Equation (6.89) can be easily inverted as

$$\eta^-(t) = \frac{RT}{\alpha^- F} \sinh^{-1}\left(\frac{j^-(t)}{2i_0^-}\right). \tag{6.90}$$

The full equation (6.89) can also be inverted to more accurately capture the effect of concentration changes on the response. The overpotential in the negative electrode is

$$\eta^-(t) = \phi_s^-(t) - \phi_e^-(t) + 0.9263 + \frac{RT}{F} \ln(c_{s,ref}), \tag{6.91}$$

where $\phi_s^-(t)$ is the solid-phase potential, $\phi_e^-(t)$ is the electrolyte-phase potential, p is the reaction order of atomic hydrogen in the metal hydride electrode, and the last two terms are the open-circuit potential. The assumption of a highly conductive electrolyte means that $\phi_e(x, t)$ is almost constant across the cell. Assuming that the electrolyte potential is uniformly zero (or ground) across the cell, ϕ_e^- drops out of Equation (6.91). In the positive electrode,

$$j^+(t) = i_0^+ \left\{ \frac{c_{s,e}^+(t)}{c_{s,ref}} \exp\left[\frac{\alpha^+ F}{RT}\eta^+(t)\right] - \left(\frac{c_{s,max} - c_{s,e}^+(t)}{c_{s,max} - c_{s,ref}}\right) \exp\left[-\frac{\alpha^+ F}{RT}\eta^+(t)\right]\right\}. \tag{6.92}$$

Again, one can invert the full equation (6.92) to solve for $\eta^+(t)$ or assume that $c_{s,e}^+ \approx c_{s,ref}$ to obtain the simplified solution

$$\eta^+(t) = \frac{RT}{\alpha^+ F} \sinh^{-1}\left(\frac{j^+(t)}{2i_0^+}\right), \tag{6.93}$$

where the overpotential

$$\eta^+(t) = \phi_s^+(t) - 0.427 - \frac{kRT}{F}\left(1 - \frac{2c_{s,e}^+(t)}{c_{s,max}}\right), \tag{6.94}$$

with the intercalation constant of the nickel electrode $k = 0.789$ and the electrolyte potential is assumed to be zero.

6.3.4 Cell Voltage

With boundary conditions applied galvanostatically, the cell current is the model input. The output voltage across the cell terminals is calculated from

$$V(t) = \phi_s^+(t) - \phi_s^-(t) - \frac{R_f}{A}I(t), \tag{6.95}$$

where R_f is an empirical contact resistance. Using the previous equations, the model simplifies to

$$\frac{C^+_{s,e}(s)}{I(s)} = \frac{G^+_{s,e}(s)}{a^+_s AL^+}, \tag{6.96}$$

$$V(t) = \frac{RT}{\alpha^+ F} \sinh^{-1}\left(\frac{j^+(t)}{2i^+_0}\right) + 1.3533 + \frac{kRT}{F}\left(1 - \frac{2c^+_{s,e}(t)}{c_{s,max}}\right)$$
$$- \frac{RT}{\alpha^- F} \sinh^{-1}\left(\frac{j^-(t)}{2i^-_0}\right) + \frac{RT}{F}\ln(c_{s,ref})^p - \frac{R_f}{A}I(t). \tag{6.97}$$

Note that the H^+ concentration in the negative electrode ($c^-_{s,e}(t)$) is not used to calculate voltage, so the transfer function (6.83) is not needed.

6.3.5 Simulation Results

To investigate the nonlinear response of the Ni–MH model, full discharge and charge curves are calculated using the simple particle models in Equations (6.85) and (6.86) and by inverting the full nonlinear Butler–Volmer equations (6.89) and (6.92). The model parameters are given in Table 6.5 from [56].

Figure 6.17 shows the simulation results for discharging the Ni–MH cell from 100% SOC ($c^+_s = 0$ and $c^-_s = c_{s,max}$). The applied discharge currents are 0.5 A, 1.0 A, and 1.5 A, corresponding to the discharge rates of $C/2$, $C/1$, and $1.5C$, respectively. The cell potential decreases gradually with time due to the decrease of the hydrogen concentration in the metal hydride and the increase of the proton concentration in the nickel active material. At the end of discharge, the nickel electrode is fully charged with $c^+_s = c_{s,max}$ while the MH electrode

Table 6.5 Parameters for Ni–MH cell model

Parameter	Negative (MH)		Positive (Ni)
Design specifications			
Thickness, δ (cm)	0.04		0.07
Particle radius, R_s (cm)	22.5×10^{-4}		2.3×10^{-4}
Electrode plate area, A (cm^2)		30	
Current collector contact resistance, R_f (Ω cm^2)		3	
Hydrogen concentrations			
Maximum solid phase concentration $c_{s,max}$ (mol/cm^3)	0.1025		0.0383
Reference concentration, $c_{s,ref}$ (mol/cm^3)	$c_{s,max}$		$0.5c_{s,max}$
Kinetic and transport properties			
Exchange current density, i_0 (A/cm^2)	3.2×10^{-4}		1.04×10^{-4}
Specific interfacial area, a_s (cm^2/cm^3)	693		4033
Charge-transfer coefficients, α		0.5	
Solid phase H$^+$ diffusion coefficient, D_s (cm^2/s)	2.0×10^{-8}		4.6×10^{-11}

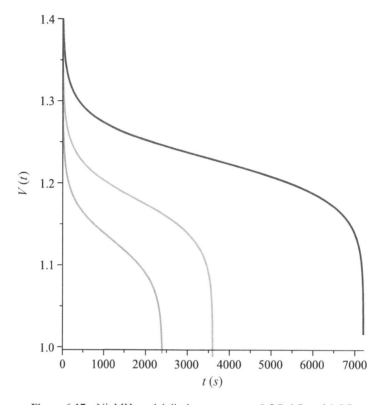

Figure 6.17   Ni–MH model discharge curves at 0.5C, 1C, and 1.5C

concentration is not depleted to zero. This is because the capacity of the nickel electrode is less than the MH electrode. In other words, the Ni–MH cell is positive electrode-limited. Ni–MH cells are designed to be positive electrode-limited to avoid hydrogen evolution in the negative electrode when the cell is fully charged and overcharged.

The simulated discharge curves exhibit the following characteristics: an initial quick drop in the cell potential, followed by a large portion of a much more shallow potential drop with a slope dependent on the discharge rate according to Equation (6.94), and ended with a drastic drop in the cell potential due to the depletion of active materials in the nickel electrode (positive electrode-limiting). The rate of discharge significantly affects the cell behavior: the larger the rate is, the more quickly the cell potential drops. The simulation results show that the cell has a capacity of 1 Ah because the 0.5C discharge takes 2 h to complete and the 1.5C discharge takes only a 0.5 h. The nonlinear shape of the discharge curve is entirely due to nonlinearities in the Butler–Volmer equations.

Figure 6.18 shows the simulation results for charging the Ni–MH cell from 0% SOC ($c_s^+ = c_{s,max}$ and $c_s^- = 0$) to fully charged. The same currents are applied as in the discharge case, but with opposite sign. During charge, the cell potential exhibits a quick rise at the beginning of charge, followed by a stage of gradually increasing potential. The larger the charge rate is, the more quickly the cell potential increases. The actual SOC, however, may not

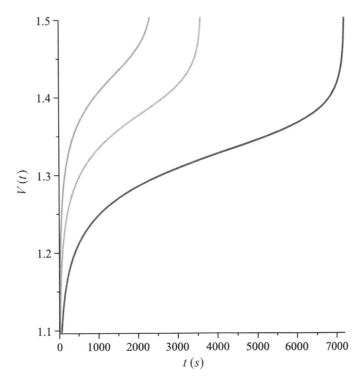

Figure 6.18   Ni–MH model charge curves at 0.5C, 1C, and 1.5C

reach 100% because charging is typically cut off when a maximum safe voltage is achieved. Again, owing to the different capacities of the two electrodes, the nickel electrode is fully charged before the metal hydride electrode. The charge curves are similar in shape to the discharge curves. Initially, the voltage rises quickly, followed by a sloping increase, and finally a sharp increase at the end of charge.

6.3.6   Linearized Model

Linearized models are easier to simulate and use for model-based estimation and control. They are often sufficiently accurate, especially if the current rates and SOC range are small. In this section, the nonlinear model is simplified to include only the nickel electrode dynamics and then linearized at a nominal SOC. The linearized and nonlinear simulation results are then compared.

The particle dynamics are distributed, so they must be discretized; but they are already linear, so linearization is not required. The second-order Padé approximation of the distributed cylindrical particle transfer function given in Equation (4.78) provides the discretized particle transfer function:

$$\frac{C_{s,e}^+(s)}{I(s)} = \frac{6(3[R_s^+]^4 s^2 + 128[R_s^+]^2 D_s^+ s + 640[D_s^+]^2)}{a_s^+ A L^+ R_s^+ Fs([R_s^+]^4 s^2 + 144[R_s^+]^2 D_s^+ s + 1920[D_s^+]^2)}. \tag{6.98}$$

The linearized, full inversion of Equation (6.92) is

$$\eta^+(t) = \eta_0^+ + \frac{R_{ct}^+}{A} I(t) - \frac{\partial \eta^+}{\partial c_{s,e}^+} \Delta c_{s,e}^+(t), \tag{6.99}$$

where

$$\eta_0^+ = \frac{RT}{\alpha^+ F} \ln \left[\sqrt{\frac{c_{s,ref}^+ \left(c_{s,max}^+ - c_{s,e0}^+ \right)}{c_{s,e0}^+ \left(c_{s,max}^+ - c_{s,ref}^+ \right)}} \right], \tag{6.100}$$

$$R_{ct}^+ = \frac{RT}{2i_0^+ \alpha^+ F a_s^+ L^+} \sqrt{\frac{\left(c_{s,max}^+ - c_{s,ref}^+ \right) c_{s,ref}^+}{\left(c_{s,max}^+ - c_{s,e0}^+ \right) c_{s,e0}^+}}, \tag{6.101}$$

$$\frac{\partial \eta^+}{\partial c_{s,e}^+} = \frac{RT c_{s,max}^+}{2\alpha^+ F c_{s,e0}^+ \left(c_{s,max}^+ - c_{s,e0}^+ \right)}, \tag{6.102}$$

and $\Delta c_{s,e}^+(t) = c_{s,e}^+(t) - c_{s,e0}^+$ is the concentration deviation from the linearization point, $c_{s,e0}^+$.

It is also interesting to monitor the SOC or depth of discharge (DOD = 1 − SOC) of the cell during charge and discharge. The DOD for the positive electrode of an Ni–MH cell is

$$\mathrm{DOD}(t) = \frac{c_{s,ave}^+(t)}{a_s^+ A L^+ c_{s,max}^+}, \tag{6.103}$$

where

$$\frac{C_{s,ave}^+(s)}{I(s)} = \frac{1}{V^+ s} \int_0^{R_s^+} C_{s,e}^+(r, s) \, dV^+ = \frac{2}{R_s^+ F s} \tag{6.104}$$

is the volumetric average of active material in the electrode. Performing the integral in Equation (6.104) yields the DOD transfer function

$$\frac{\mathrm{DOD}(s)}{I(s)} = \frac{2}{a_s^+ A L^+ R_s^+ F c_{s,max}^+ s}. \tag{6.105}$$

Figure 6.19 shows the response of the positive electrode to a pulse charge/discharge input current. The cell capacity has a maximum discharge of 5 A (5C rate) and maximum charge of 2 A (2C rate). The cycle has more discharge than charge, so the SOC shown in the top figure slowly decreases over time. The rate of SOC change is directly proportional to the current rate. The H^+ concentration is increasing in the positive electrode, caused by the flow of electrons into the electrode associated with cell discharge. An exponential response in the H^+ concentration is observed due to the diffusion transport through the particle. The voltage decays as the battery is being discharged. For short times with the SOC near the operating point and relatively low currents, the linearized model closely approximates the nonlinear response. As time marches on, however, and the SOC begins to deviate significantly from the linearization point, the match is not as accurate, especially during the high current (5C)

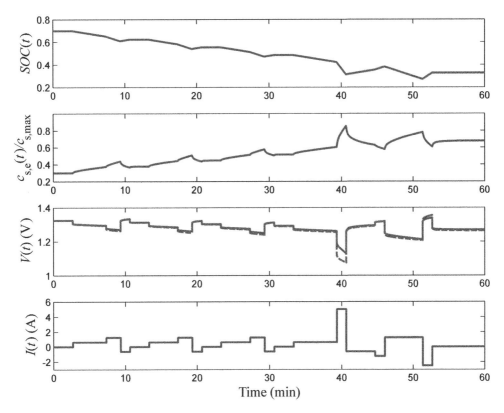

Figure 6.19 Pulse charge/discharge response of an Ni–MH cell: nonlinear (solid) and linearized (dashed)

discharge around 40 s. These results indicate that a linearized model has the potential to accurately match the nonlinear response without the numerical and analytical complexity associated with the nonlinearities.

Problems

6.1 Assuming zero electrolyte diffusion and a single electrode particle in a single electrode, develop and simulate a linear model of an Li-ion cell. The transcendental transfer function for Li$^+$ concentration on the surface of the particle is given by the transcendental transfer function

$$\frac{C_{\mathrm{s,e}}(s)}{J(s)} = \frac{1}{a_{\mathrm{s}}F}\left(\frac{R_{\mathrm{s}}}{D_{\mathrm{s}}}\right)\left[\frac{\tanh(\beta)}{\tanh(\beta) - \beta}\right], \tag{A.6.1}$$

where $\beta = R_{\mathrm{s}}\sqrt{s/D_{\mathrm{s}}}$. The output voltage is given by

$$V(t) = \eta(t) + U(c_{\mathrm{s,e}}(t)), \tag{A.6.2}$$

where the linearized Butler–Volmer equation

$$\frac{\eta(s)}{I(s)} = \frac{R_{ct}}{a_s} \frac{J(s)}{I(s)} \tag{A.6.3}$$

with

$$J(s) = -\frac{I(s)}{A\delta}. \tag{A.6.4}$$

(a) Using a second-order Padé approximation of the transcendental transfer function (A.6.1), derive a third-order transfer function $V(s)/I(s)$ in closed form.
(b) Calculate and plot the frequency response and compare with Figure 6.10.
(c) Simulate and plot the time response to the pulse input in Figure 6.11 and compare the results.
(d) How well does this simplified model agree with the more complex model presented in this chapter? What can the more complicated model predict that the simplified model in this problem cannot?

The parameters for the dominant, positive electrode are $a_s = 1.5 \times 10^4$/cm, $R_s = 1 \times 10^{-4}$ cm, $D_s = 3.7 \times 10^{-12}$ cm^2/s, $U(c_{s,e}(t)) = 3.6 - 71c_{s,e}(t)$ V, $R_{ct} = 9.71$ Ω cm^2, $\delta = 37 \times 10^{-4}$ cm and $A = 10\,452$ cm^2. Note: use these parameter values only for parts (b)–(d).

6.2 Augment the model in Problem 6.1 with the second, negative electrode, again neglecting electrolyte diffusion.

(a) Using a second-order Padé approximation of the transcendental transfer function (A.6.1) for this electrode as well, derive a fifth-order transfer function $V(s)/I(s)$ in closed form.
(b) Calculate and plot the frequency response and compare with Figure 6.10.
(c) Simulate and plot the time response to the pulse input in Figure 6.11 and compare the results.

The parameters for the negative electrode are $a_s = 1.74 \times 10^4$/cm, $R_s = 1 \times 10^{-4}$ cm, $D_s = 2 \times 10^{-12}$ cm^2/s, $U(c_{s,e}(t)) = 0.103 - 3.9c_{s,e}(t)$ V, $R_{ct} = 7.01$ Ω cm^2, and $\delta = 50 \times 10^{-4}$ cm. Note: use these parameter values only for parts (b) and (c).

6.3 Augment the model in Problem 6.1 with electrolyte diffusion by assuming that the Li$^+$ concentration is symmetric about the center of the cell, giving the boundary condition

$$\left.\frac{\partial c_e}{\partial x}\right|_L = -\alpha c_e(L, t), \tag{A.6.5}$$

where $\alpha = 2/\delta_{sep}$. The other boundary condition is $\frac{\partial c_e}{\partial x}(0, t) = 0$. The field equation, assuming constant current density, is

$$\frac{\partial c_e}{\partial t} = a_1 \frac{\partial^2 c_e}{\partial x^2} + a_2 j(t). \tag{A.6.6}$$

The electrolyte concentration couples into the cell voltage through conservation of charge in the electrolyte. Assume that this coupling changes the output voltage equation to

$$V(t) = \eta(t) + U(c_{s,e}(t)) + a_3 c_{e,avg}(t),\tag{A.6.7}$$

where

$$c_{e,avg}(t) = \frac{1}{L}\int_0^L c_e(x,t)\,dx.\tag{A.6.8}$$

(a) Using a second-order Padé approximation of the transcendental transfer function (A.6.1), derive a fourth-order transfer function $V(s)/I(s)$ in closed form.
(b) Calculate and plot the frequency response and compare with Figure 6.10.
(c) Simulate and plot the time response to the pulse input in Figure 6.11 and compare the results.

The additional parameters are $a_1 = 1.84 \times 10^{-6}$ cm^2/s, $\delta_{sep} = 25.4 \times 10^{-4}$ cm, $a_2 = 1.32 \times 10^{-5}$ mol/C, and $a_3 = 0.033$ V cm^3/mol. Note: use these parameters only for parts (b) and (c).

6.4   Augment the model in Problem 6.1 with the nonlinear Butler–Volmer equation

$$j = i_0\left[\exp\left(\frac{\alpha_a F}{RT}\eta\right) - \exp\left(-\frac{\alpha_c F}{RT}\eta\right)\right],\tag{A.6.9}$$

where

$$i_0 = k(c_e)^{\alpha_a}(c_{s,max} - c_{s,e})^{\alpha_a}(c_{s,e})^{\alpha_c}.\tag{A.6.10}$$

The system parameters are $\alpha_a = \alpha_c = 0.5$, $k = 6.3714$ A/cm^2, $c_{s,max} = 23.9 \times 10^{-3}$ mol/cm^3, and $i_0 = 3.6 \times 10^{-3}$ A/cm^2. Assume that the electrolyte concentration is constant at $c_e = 1.2 \times 10^{-3}$ mol/cm^3.

(a) Develop a nonlinear simulation model of this system and print out the block diagram.
(b) Simulate and plot the time response to the pulse input in Figure 6.11 and compare the results.
(c) Simulate and plot the time response to the pulse input in Figure 6.11 multiplied by five times. Compare the results in parts (b) and (c).
(d) Simulate and plot the time response to a constant $1C$, $2C$, and $5C$ discharge from 100% SOC to 0% SOC.

6.5   For the Pb–acid cell shown in Figure 6.1, average the electrolyte diffusion in Equation (6.3) to obtain a differential equation for the average acid concentration $c_{avg}(t)$, assuming that the electrolyte potential is zero and the porosity ε is constant. Use Equations (6.9)–(6.10), assuming $c(x,t) = c_{avg}(t)$, to obtain a third-order, linear, state-space model. The model input is

$$\mathbf{u} = \begin{bmatrix} I(t) \\ U_{PbO_2} \end{bmatrix} \quad \text{and the state vector is} \quad \mathbf{x} = \begin{bmatrix} c_{avg} \\ \phi_{sp} \\ \phi_{sm} \end{bmatrix}.\tag{A.6.11}$$

The output voltage $V(t) = \phi_{sp} - \phi_{sm}$.

(a) Derive the state matrices in closed form for the third-order Pb–acid cell model.

(b) Simulate and plot the time response to the $0.1C$ charge and discharge shown in Figure 6.4.

(c) Simulate and plot the response to the switcher cycle input in Figure 6.8. Use the parameters for charge.

(d) How well does this simplified model agree with the more complex model presented in Chapter 6? What can the more complicated model predict that the simplified model in this problem cannot?

Parameters R_a, a_{dl}, C_{dl} have different values in the positive and negative electrode, as shown in Table 6.1. Note: use these parameters for parts (b)–(d) only.

7

Estimation

Estimation and prediction play important roles in providing feedback to the user and as the basis for advanced BMSs. In this chapter we introduce methods for estimating the SOC and state of health (SOH) of batteries.

Following [10] and [60], SOC is defined as the percentage of the maximum possible charge that is present inside a rechargeable battery. SOC is not directly measurable in a battery, so it must be inferred from other measurements or estimated in some way. Voltage, current, and sometimes temperature are measured for each battery in a pack. From this information, the SOC estimate is constructed in real time and provided to the user and to the BMS.

Although accurate SOC estimation is desirable for portable electronics, it is the demanding HEV application that motivates the development of high-fidelity SOC estimators. In portable electronics, the SOC estimator provides information to the user that warns of impending power loss. The exact SOC at any particular instant of time is not of critical importance to the user. The power demands on the battery are typically modest and the SOC slowly decreases over time. In HEV application, however, the SOC changes quickly and by large amounts due to rapid acceleration and hard regenerative braking. Accurate SOC estimation is critical for high-performance HEV BMSs. Charge and discharge constraints can depend on SOC.

For further information on SOC estimation, interested readers are referred to the review articles by Piller *et al.* [61] and Pop *et al.* [62]. Plett pioneered the use of Kalman filtering for equivalent-circuit models of batteries [60, 63, 64]. Santhangopalan and White used extended Kalman and unscented filters with physics-based models to predict SOC in Li-ion cells [59, 65, 66]. Smith *et al.* [67] and Di Domenico *et al.* [68] investigated different model-based SOC estimators using Kalman filters. Wang *et al.* [69] presented an equivalent-circuit-based SOC estimator.

As batteries age, their capacity decreases and internal impedance increases. The degradation mechanism depends on the battery chemistry, manufacturer, and design. Often, different degradation mechanisms operate at different stages in the battery's life. It can be difficult to differentiate between environmental (e.g., low or high temperature) and aging effects. The SOH is defined as the battery capacity divided by the initial capacity. After the required break-in period where the capacity can increase, the battery is at 100% SOH. As the battery ages, the SOH steadily decreases until the battery can no longer be used and it has reached its end of life (EOL), typically around 80% SOH [70, 71]. The energy stored in the battery depends

Battery Systems Engineering, First Edition. Christopher D. Rahn and Chao-Yang Wang.
© 2013 John Wiley & Sons, Ltd. Published 2013 by John Wiley & Sons, Ltd.

on both the SOC and the SOH. Accurate prediction of SOH is needed in addition to accurate SOC estimation to provide users with reliable information on the remaining run time without charging. Advanced BMSs can use the SOH information to change the control strategy over the life of the battery, maximizing its performance and longevity.

Spotnitz [72] reviewed the major characteristics of capacity fade in Li-ion batteries: impedance growth and capacity loss. The available capacity fade data and characteristics are summarized. Wenzl et al. [73] reviewed the methods that can be used for lifetime prediction and summarized the available techniques for SOH estimation. Coleman et al. [74] used a two-pulse test to estimate SOC and SOH of VRLA batteries. A variety of other methods are cited, including a full discharge test, internal resistance test, and impedance method. The last two methods and the two-pulse test essentially measure the high-frequency impedance of the cell. Plett [75] used total least squares on current integration and SOC estimate signals to best fit SOH. Safari et al. [71] applied mechanical-fatigue life-prognostic methods to predict the life of Li-ion batteries. The damage accumulated for one cycle or interval of operation adds to the damage from previous cycles so the SOH estimation consists of cycle counting and comparing with the total number of cycles at EOL. There are also many patents on capacity estimation and SOH (e.g., see Plett [76] and cited references).

The models developed, discretized, and simulated in the previous chapters provide an excellent basis for SOC and SOH estimation. The focus of this chapter is on model-based estimation, but non-model-based approaches of SOC and SOH estimation will also be discussed. The intellectual effort required to derive and solve the model yields a greater understanding of the underlying processes and a physical feel for the battery dynamics. The model can be used to motivate and evaluate model-free estimation methods or it can be incorporated into a model-based estimation framework.

7.1   State of Charge Estimation

To provide a precise definition of SOC, we first define the nominal capacity of a cell C to be the maximum number of ampere-hours that can be drawn from the fully charged cell at room temperature and a $C/30$ rate. Capacity can also be expressed in units of coulombs. The remaining capacity $C_r(t)$ is defined as the number of ampere-hours that can be drawn from the cell starting from the current time t, at room temperature, and at a $C/30$ rate. We define

$$\text{SOC} = \frac{C_r(t)}{C} = 1 - \frac{1}{C} \int_0^t I(\tau) \, d\tau, \tag{7.1}$$

assuming the initial SOC at $t = 0$ is 100% and $I(t)$ is the applied current with $I > 0$ during discharge.

To provide an electrical analogy for battery capacity, we determine the capacity of a capacitor. A fully charged capacitor holds a maximum charge $Q_{max} = C_e V_{max}$, where C_e is the capacitance in farads and V_{max} is the maximum voltage. The maximum stored charge Q_{max} is the capacity of the capacitor. The SOC of a capacitor is

$$\text{SOC} = \frac{Q(t)}{Q_{max}} = \frac{C_e V(t)}{C_e V_{max}} = \frac{-1}{C_e V_{max}} \int_0^t I(\tau) \, d\tau, \tag{7.2}$$

where $\text{SOS}(0) = 0\%$ and $I < 0$ during charge.

The voltage on a capacitor increases linearly with charge, so the energy stored in the capacitor is

$$E = \int_0^Q V \, dq = \int_0^Q \frac{q}{C_e} \, dq = \frac{1}{2} \frac{Q^2}{C_e} = \frac{1}{2} \frac{Q_{max}^2 SOC^2}{C_e}. \tag{7.3}$$

Thus, the stored energy of a capacitor is proportional to the square of the SOC.

Battery cell voltages also depend on SOC; so, as with capacitors, battery SOC does not measure energy storage. Cell voltage decreases with decreasing SOC, initially at a low slope and then more quickly as DOD (DOD $= 1 -$ SOC) reaches one. Some chemistries (e.g., lithium iron phosphate) have a very flat voltage versus DOD curve until large DOD, where the curve drops sharply. Others (e.g., PbC) have a fairly linear voltage versus DOD curve, similar to capacitors. All cells display some decrease in voltage with DOD.

As with capacitors, the capacity of a cell depends on the operating voltage range. A cell is fully charged when it maintains the prescribed cell voltage for its chemistry V_h (e.g., 4.2 V for an Li-ion cell) in steady state and at room temperature. This can be achieved through, for example, trickle charging at V_h for a suitably long period of time. The V_h voltage is limited by side reactions that can damage and reduce the life of the battery at high voltages. A fully discharged cell that provides the maximum number of ampere-hours from a full charge has a steady-state voltage of V_l (e.g., 3.0 V for an Li-ion cell) at room temperature. The V_l voltage is also chosen to limit battery damage. The voltage versus DOD curve is steeply sloped at high DOD, so operating at a reduced V_l may not result in a significant increase in stored energy.

High discharge rates, low temperatures, and aging can significantly reduce battery capacity. A battery at 80% SOC may only be able to provide 20% of the rated ampere-hours at $5C$ discharge rate compared with $C/30$ because the battery voltage drops quickly to V_l before the rated capacity is achieved. If the high-rate discharge stops, however, the voltage can recover and further capacity can be removed from the battery. Old batteries, or those operating at low temperatures, have nominal capacities that can be significantly less than the nominal capacity of the fresh cell at room temperature. Again, an SOC of 80% for an aged or cold battery may mean that only 20% of the room temperature and fresh cell capacity is available to be discharged.

From a systems perspective, SOC mirrors the gas gauge in a traditional vehicle powered by an ICE. The distance one can travel on a quarter tank of gas, or range, depends on the rate at which fuel is consumed. In an ideal situation, speed is proportional to fuel consumption rate. In practice, however, range estimation is complicated by many unknown factors. The weight of the vehicle and road grade are not known a priori. The efficiency of the ICE and aerodynamic drag depend on vehicle speed. The efficiency drops to zero at stop lights or traffic jams. Unlike battery SOC, however, the gas gauge measures the energy remaining in the tank.

Many of the decisions made by an HEV control system depend on SOC. In real time, the BMS must decide whether to use mechanical brakes or regenerative braking during deceleration or, during acceleration, draw power from the battery pack or the ICE. These decisions are often based on the SOC. Unlike the gas gauge, however, there is no sensor that directly measures SOC, so an estimate must be used. The accuracy of this estimate is critical to the proper, safe, and efficient performance of the HEV.

In most applications, the SOC estimator must be as non-intrusive as possible. In this section, we assume that only the voltage, current, and temperature are measured. Individual cell voltages and temperatures are rarely measured, and some packs may not measure the voltages and temperatures of individual batteries. Batteries and packs can be formed from series and parallel connections of cells and batteries, respectively. The SOCs for batteries and packs, however, are averaged over the individual cell and battery SOCs, respectively. We also assume that the BMS uses charge and discharge current only to fulfill the application requirements. The SOC estimation scheme cannot command the battery current and introduce pulses or sine sweeps in the battery current to aid in SOC estimation. This would require complicated and expensive equipment and limit the availability of the battery for its intended purpose. Finally, we assume that the sensors are sampled at a finite bandwidth, typically around 10 Hz. Thus, high-frequency dynamics need not be modeled and cannot be relied upon to help with SOC estimation.

With an SOC of 100% the battery can provide $C/30$ amps for 30 h. This does not necessarily mean that the battery can provide $2C$ amps for 0.5 h. The diffusion processes within the cell create nonuniform concentration and potential distributions. The ions can only travel so fast through the cell, and if the current draw is too high then the voltage drops precipitously, resulting in a sudden loss of power. It is as if the fuel in the gas tank of an ICE vehicle is flowing through a very small tube and is rate limited. The rate-limiting process in an ICE is typically not fuel supply, however, but combustion. Increasing the fuel flow rate above a rate-limited value (depending on displacement, compression ratio, fuel injection, etc.) does not increase the power output but simply floods the engine. The rate-limiting characteristics of batteries mean that their design and integration into the powertrain are more critical and complex than a simple fuel tank.

In this section we study three methods that are used for SOC estimation in batteries: voltage lookup, current counting, and state estimation (Luenberger observer). The exact method used for a given battery can involve combinations of these methods and other empirical techniques.

7.1.1 SOC Modeling

In order to develop more advanced estimators, SOC must be related to the physical processes going on inside the battery cell. We would like to be able to determine the SOC of a cell simply by measuring the current state without knowing the past history. SOC calculated from Equation (7.1) requires the complete past time history of $I(t)$ and an accurate SOC value at $t = 0$. The equations that relate SOC to battery state are critical for model-based estimation and control.

Physically, SOC measures the quantity of active materials remaining in the cell. The chemistry in the anode and cathode converts materials from one form to another. As current is discharged from the battery, active material is converted to inactive material. Eventually, all of the active material is used up and no further current can be produced.

The material that is consumed inside the cell depends on the battery chemistry. For Li-ion batteries, lithium ions move from the negative electrode to the positive electrode during discharge. The negative electrode becomes depleted of lithium and the positive electrode becomes saturated. Eventually, either the negative electrode can no longer supply lithium ions and/or the positive electrode can no longer hold lithium ions and current stops flowing. In a well-designed battery, the anode and cathode are balanced so that both can absorb the same amount of lithium ions. Often the anode is slightly oversized (e.g., by 10%), however, to limit an aging mechanism that is predominant in that electrode during undercharge or overcharge. For Li-ion batteries, the average lithium concentration is proportional to SOC.

Using the Li-ion model from Chapter 6, we can generate an explicit formula for SOC and the battery capacity C of Li-ion cells as an example. Conservation of lithium in the spherical material particles is governed by

$$\dot{c}_s = \frac{D_s}{r^2} \left[r^2 c'_s \right]' \tag{7.4}$$

with boundary conditions

$$c'_s(0, t) = 0 \quad \text{and} \quad c'_s(R_s, t) = -\frac{j R_s}{3 \varepsilon_s F}. \tag{7.5}$$

The volume-averaged lithium concentration is

$$c_{s_{avg}} = \frac{1}{V_s} \int c_s \, dV_s, \tag{7.6}$$

where $V_s = 4\pi R_s^3 / 3$ and $dV_s = 4\pi r^2 \, dr$. Volume integration of Equation (7.4) yields

$$\dot{c}_{s_{avg}} = \frac{3 D_s}{R_s} \left[R_s^2 c'_s(R_s, t) \right] = -\frac{1}{\varepsilon_s F} j_{avg} \tag{7.7}$$

using the boundary conditions (7.5). The average current density is obtained by averaging the conservation of charge equation

$$j_{avg} = \begin{cases} \int_0^{\delta_m} [\sigma^{eff} \phi'_s]' \, dx = \dfrac{1}{\delta_m A} I & \text{for the negative electrode,} \\[2mm] -\int_{\delta_m + \delta_{sep}}^{L} [\sigma^{eff} \phi'_s]' \, dx = -\dfrac{1}{\delta_p A} I & \text{for the positive electrode,} \end{cases} \tag{7.8}$$

using the zero-flux boundary conditions at the separator and $-\sigma_m^{eff} A \phi'_s(0, t) = \sigma_p^{eff} A \phi'_s(L, t) = I$. Substitution of Equation (7.8) into Equation (7.7) yields the average concentration dynamics in the negative electrode,

$$\dot{c}_{s_{avg}}^- = -\frac{1}{\delta_m A \varepsilon_m F} I, \tag{7.9}$$

and positive electrode,

$$\dot{c}_{s_{avg}}^+ = \frac{1}{\delta_p A \varepsilon_p F} I. \tag{7.10}$$

Equations (7.9) and (7.10) show that the average concentration in the negative and positive electrodes decrease and increase, respectively, during discharge.

SOC can be defined for either electrode and for the overall cell. Concentration is nondimensionalized by introducing the stoichiometry

$$\theta = \frac{c_{S_{avg}}}{c_{S_{max}}}, \tag{7.11}$$

where the 100% ($\theta_{100\%}$) and 0% ($\theta_{0\%}$) reference stoichiometries are determined experimentally. SOC is now defined as

$$SOC = \frac{\dfrac{c_{S_{avg}}}{c_{S_{max}}} - \theta_{0\%}}{\theta_{100\%} - \theta_{0\%}} \tag{7.12}$$

for the negative electrode, positive electrode, or average of the two to get SOC for the whole cell. SOC defined in Equation (7.12) satisfies Equation (7.1) with

$$C = \delta A \varepsilon F c_{S_{max}} [\theta_{100\%} - \theta_{0\%}]. \tag{7.13}$$

With a few simplifying assumptions, SOC can be related to the output voltage. If the electrolyte impedance is neglected then the overpotential in the negative electrode can be written

$$\eta_{avg}^- = \phi_{S_{avg}}^- - U_{avg}^- = \frac{R_{ct}^- R_S^-}{3\varepsilon_m} I, \tag{7.14}$$

where

$$U_{avg}^- = \frac{1}{\delta_m} \int_0^{\delta_m} U_m(c_{s,e}(x,t)) \, dx \approx U_m(c_{S_{avg}}^-). \tag{7.15}$$

In Equation (7.15) the surface concentration is assumed to be equal to the average concentration and uniform throughout the electrode. This is only a reasonable assumption for very low current charge/discharge. After a similar calculation for the positive electrode, the output voltage

$$V(t) \approx V_{avg}(t) = U_p(c_{S_{avg}}^+) - U_m(c_{S_{avg}}^-) - \frac{R_T}{A} I, \tag{7.16}$$

where

$$R_T = R_f + \frac{R_{ct}^+ R_S^+}{3\varepsilon_p \delta_p} + \frac{R_{ct}^- R_S^-}{3\varepsilon_m \delta_m}. \tag{7.17}$$

A similar SOC analysis can be performed for Ni–MH cells, but Pb–acid batteries consume both acid and active material during discharge, complicating the SOC analysis. Acid is consumed in the reactions with Pb and PbO$_2$ in the positive and negative plates. If the acid concentration reduces to zero, the reaction can no longer occur and the voltage drops quickly. These reactions also convert the active Pb and PbO_2 to inactive forms. Once there is no active material left to convert, the reaction stops and voltage drops. A properly designed Pb–acid cell is balanced so that acid and active material are present in proper ratios to ensure equal utilization. For Pb–acid batteries the SOC can depend on either the average acid concentration or the average active material in each electrode. In practice, one electrode and one reaction will dominate SOC for a given cell.

7.1.2   Instantaneous SOC

SOC estimation is typically based on low-order models like those introduced in the previous section. The averaging and other assumptions required to generate those models provide a reasonable estimate of the average remaining charge in the cell. The remaining charge is only half the story, however. Cell voltage plays an important practical role in the utilization of the charge stored in the cell. With low current rate discharge the voltage drops slowly and all of the charge can be withdrawn from the cell at a reasonable voltage level. Under high-rate discharge the voltage drops precipitously and the power available from the cell is greatly reduced. If the high current discharge is cut off then the voltage eventually returns to its nominal value and power can again be supplied. The high-rate discharge did not reduce the SOC to zero but only the instantaneous SOC (iSOC).

The iSOC measures the potential for rapid voltage drop in the cell. For Li-ion cells, the voltage developed depends on the concentration of lithium ions on the surface of the material particles. If the concentration drops due to diffusion limitations at any location in the electrode there can be essentially a short at that point. The solid phase conductivity is typically large, so this short can result in a rapid drop in voltage regardless of where (e.g., near the separator or current conductor) it occurs.

A simple approximate model of iSOC for Li-ion cells can be based on the averaged Equations (7.9), (7.10), and (7.16) and an extra equation that predicts the average surface concentration of lithium, $c_{s,e_{avg}}$. The analytical transfer function was derived in Chapter 4 for the material particles as

$$\frac{C_{s,e}(s)}{J_{avg}(s)} = \frac{R_s \tanh(\Gamma(s))}{3 D_s \varepsilon_s F \tanh(\Gamma(s)) - \Gamma(s)}, \tag{7.18}$$

where $\Gamma(s) = R_s \sqrt{s/D_s}$ and we assume that $J(x,s) = J_{avg}(s)$ is uniform across each electrode. Subtracting the average concentration in Equation (7.7) from Equation (7.18) yields a transfer function with a finite gain at DC:

$$\lim_{s \to 0} \left(\frac{C_{s,e}(s)}{J_{avg}(s)} - \frac{1}{s \varepsilon_s F} \right) = -\frac{R_s^2}{15 \varepsilon_s D_s F}. \tag{7.19}$$

As a first approximation, we take only the steady-state value from Equation (7.19) and neglect the rest of the particle dynamics. Thus, the average particle surface concentration can be expressed as

$$c_{s,e_{avg}} = c_{s_{avg}} - \frac{R_s^2}{15\varepsilon_s D_s F} j_{avg}. \tag{7.20}$$

This quantity has also been called the critical surface charge [68] or average surface stoichiometry [67].

Thus, the iSOC can be approximated using Equation (7.20) as follows:

$$\mathrm{iSOC} = \frac{\frac{c_{s,e_{avg}}}{c_{s_{max}}} - \theta_{0\%}}{\theta_{100\%} - \theta_{0\%}}. \tag{7.21}$$

Equation (7.21) shows that iSOC differs from SOC by the current rate because j_{avg} depends on I according to Equation (7.8). The iSOC reflects the instantaneous surface concentration resulting from current flow rather than just the average state of the electrode.

One can also incorporate Butler–Volmer and open-circuit potential nonlinearities in this model [59, 65]. This is the SP model and that was derived for Ni–MH cells in Chapter 6.

7.1.3   Current Counting Method

The current counting method is motivated by the definition of SOC in Equation (7.1) and the model equations Equations (7.9)–(7.12). If the initial SOC is known then the current SOC can be calculated by counting or integrating the current as follows:

$$\mathrm{SOC}_{cc} = \mathrm{SOC}_{cc}(0) - \frac{1}{C} \int_0^t I(\tau) \, d\tau. \tag{7.22}$$

This approach suffers from two limitations. First, one must know the initial SOC ($\mathrm{SOC}_{cc}(0)$). This can be obtained from a fully charged and rested battery. Second, noise in the current sensor can cause SOC drift. There is no feedback in Equation (7.22) to correct for drift until a known SOC is reached and $\mathrm{SOC}_{cc}(0)$ can be reset. For HEV applications, however, where the cell is continually being charged and discharged, it may be a long time between resets and the corresponding SOC_{cc} estimate may be far from the actual value.

Figure 7.1 shows the actual SOC and current counting estimate SOC_{cc} for the Ni–MH model developed in Chapter 6. The current input is primarily discharge, so the SOC decreases over the 6 min simulation. The initial SOC is assumed to be imprecisely known, so SOC_{cc} starts out at 69% instead of the actual value of 70%. The SOC estimate accurately follows the SOC trend but maintains a fixed error of 1% during the entire simulation. There is no feedback in the SOC_{cc} estimate so it cannot correct for initiation errors, drift, and sensor bias. The small-amplitude, zero-mean noise injected into the sensed current signal is effectively filtered by the integrator in the SOC estimator, however, and seems to have little effect on the SOC estimate. If an offset or gain error in the sensor is introduced, however, SOC_{cc} would drift away from the actual estimate. At some point, the cell must be fully charged to reset $\mathrm{SOC}_{cc} = 100\%$. The

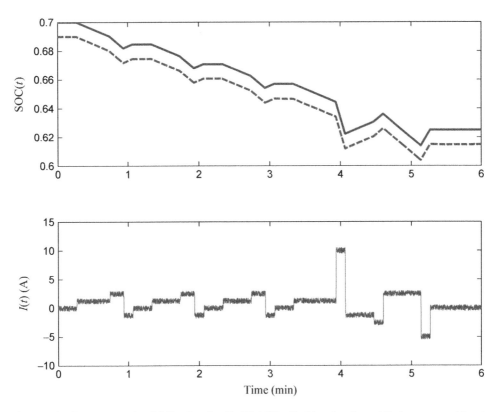

Figure 7.1 Current counting SOC estimation for Ni–MH cell with pulse charge/discharge current input: actual (solid) and estimated (dashed)

frequency of resetting depends on the desired SOC accuracy and the magnitude of the current sensor noise/drift.

7.1.4 Voltage Lookup Method

The voltage lookup method is motivated by Equation (7.16), which relates the cell output voltage $V(t)$ to the average lithium concentration. From Equation (7.12), SOC is proportional to average lithium concentration. Thus, if one simply measures the output cell voltage and compares it with a lookup table of open-circuit voltage versus SOC, one can obtain the SOC_{vl} estimate of SOC. The actual voltage signal, however, responds to the iSOC much more than SOC, since the output voltage depends more on the surface concentration than on the average concentration. The actual voltage signal also includes the contact resistance, Butler-Volmer resistance, and diffusional dynamics. It may be possible to estimate the ohmic resistance and subtract the IR voltage from the measured value. SOC_{vl} is often filtered to eliminate the effects of short-term voltage swings resulting from unmodeled dynamics. This may lead to unacceptable delays between the actual SOC and SOC_{vl}. Pop *et al.* [77] detail the practical limitations of voltage lookup for SOC estimation.

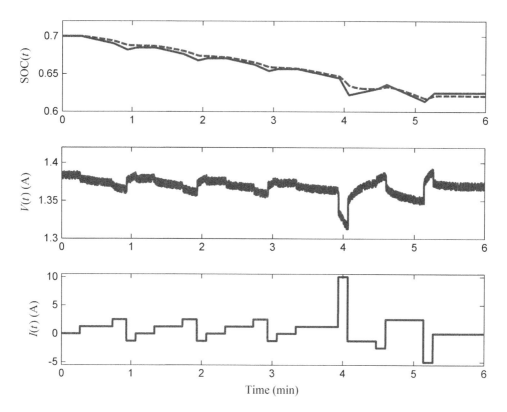

Figure 7.2 Voltage lookup SOC estimation for Ni–MH cell with pulse charge/discharge current input: actual (solid) and estimated (dashed)

Figure 7.2 shows the performance of filtered voltage lookup for SOC estimation. Small-amplitude, zero-mean noise has been added to the voltage signal to simulate the sensor accuracy in practice. At each time step, the SOC corresponding to the measured voltage is calculated and passed through a filter with a time constant of 200 s. The filter does a good job of eliminating spikes associated with current pulses but introduces delay in SOC_{vl}. The maximum SOC errors still occur at the current spikes, however; but instead of overshooting, SOC_{vl} underpredicts the change in SOC. The time constant can be tuned to minimize overshoot/undershoot in the SOC estimate, but the error cannot be eliminated entirely.

7.1.5 State Estimation

The state estimation method formalizes the concept of filtering the measured voltage to estimate SOC as described in the previous section. A model-based state estimator or Luenberger observer is designed with the primary output being SOC_{lo}. One advantage of using an observer-based approach is that it can more rigorously account for noise in the sensors by using a linear quadratic Gaussian (LQG) algorithm. Second, it can estimate more than simply SOC_{se}, such as concentration or potential distributions throughout the cell, if the underlying model is sufficiently sophisticated. Increased model sophistication, however, results in higher order estimators that are more difficult and costly to implement. In this section, state estimators are derived for a low-order averaged model similar to what was derived in Section 7.1.1 and a higher order model that predicts surface concentration on the material particles.

Observability

In order to successfully implement an observer, the underlying model must be observable from the single measured output: cell voltage. In other words, given the known current input, the observer must be able to uniquely determine the state from the sensed output. Two factors limit the observability of battery models. First, the processes in the anode and cathode are often sufficiently similar and combine in such a way that the states of the two electrodes cannot be uniquely determined. Second, if the voltage dependency on concentration is flat then the system is unobservable.

Consider the observability of the Li-ion averaged model developed in Section 7.1.1. The linearized model can be written in state-variable form as follows:

$$\dot{\mathbf{x}} = \mathbf{A}\mathbf{x} + \mathbf{b}I, \tag{7.23}$$

$$V = \mathbf{c}^T\mathbf{x} + dI, \tag{7.24}$$

where the state $\mathbf{x} = [c_{S_{avg}}^-, c_{S_{avg}}^+]^T$,

$$\mathbf{A} = \begin{bmatrix} 0 & 0 \\ 0 & 0 \end{bmatrix}, \quad \mathbf{b} = \begin{bmatrix} -\dfrac{1}{\delta_m A \varepsilon_m F} \\ \dfrac{1}{\delta_p A \varepsilon_p F} \end{bmatrix}, \quad \mathbf{c} = \begin{bmatrix} -\dfrac{\partial U_m}{\partial c} \\ \dfrac{\partial U_p}{\partial c} \end{bmatrix}, \tag{7.25}$$

and $d = -R_T/A$. The partial derivatives in $\mathbf{c}$ are evaluated at the linearization point corresponding to a specific SOC.

The observability of a state variable system is governed by the observability matrix

$$\mathcal{O} = \begin{bmatrix} \mathbf{C} \\ \mathbf{CA} \\ \vdots \\ \mathbf{CA}^{N-1} \end{bmatrix} \tag{7.26}$$

where $\mathbf{A}$ is an $N \times N$ matrix [39]. If the rank($\mathcal{O}$) $= N$ then the system is observable. The observability matrix for the state-variable system in Equations (7.23) and (7.24) is

$$\mathcal{O} = \begin{bmatrix} -\dfrac{\partial U_m}{\partial c} & \dfrac{\partial U_p}{\partial c} \\ 0 & 0 \end{bmatrix}. \tag{7.27}$$

The rank($\mathcal{O}$) $= 1 < N = 2$, so this system is unobservable.

There are two ways to circumvent the observability problems associated with two-electrode cell models. First, the two integrators can be lumped into a single state that integrates the flow of ions from one electrode to the other. This assumes that the two electrodes are balanced and ions are not consumed in side reactions in one electrode or the other. Second, one can insert a reference electrode in the separator between the anode and cathode. This allows the sensing of the voltage generated by both electrodes and greatly improves the observability of the system.

Even if the two integrators are lumped into a single state, the voltage versus concentration curve cannot be flat at the linearization point or the system will again be unobservable. A flat curve has $\partial U/\partial c = 0$, making voltage independent of concentration. A reference electrode can also address this problem if the overall voltage curve is flat, but one electrode has a nonzero slope.

Observer Design

State estimators use the input current and output voltage signals from the cell to reproduce internal signals that are not directly measurable (e.g., SOC). The state includes the internal signals of interest and other variables that change during charge and discharge. Model complexity is reflected by the number of states. Higher order models with more states result in higher order estimators that are more complex to design and implement. Estimators are implemented in real time, meaning that the estimate is updated by code running on a microprocessor on board the battery-powered device. Here, the focus is on continuous time estimators that require integration of continuous time signals. Implementation, however, requires conversion of the continuous time estimator to discrete time code that runs on the microprocessor because the signals are sampled at a fixed rate and estimate propagation takes some time. Fortunately, this is a fairly straightforward conversion if the sample time is sufficiently fast. Modern microprocessors have no problem implementing estimators with tens of states at thousands of hertz sample rates. For a battery pack with a large number of cells, however, the estimation problem can rapidly overwhelm a microprocessor. It is important, therefore, to use the lowest order estimator that provides the requisite estimation accuracy.

Figure 7.3 shows a block diagram of a Luenberger observer for cell state estimation. The cell dynamics are represented in state-variable form:

$$\dot{\mathbf{x}} = \mathbf{A}\mathbf{x} + \mathbf{b}I(t), \tag{7.28}$$

$$V(t) = \mathbf{c}^{\mathrm{T}}\mathbf{x} + dI(t), \tag{7.29}$$

with input current $I(t)$ and output voltage $V(t)$. The state matrices $(\mathbf{A}, \mathbf{b}, \mathbf{c}, d)$ and state $\mathbf{x} \in \mathcal{R}^N$ can be determined from the simplified, averaged SOC dynamics or more complicated models, like those developed in Chapter 6. All models that predict SOC have an integrator state that is proportional to SOC. The SOC output must be defined such that $\mathrm{SOC}(t) = \mathbf{c}_{\mathrm{SOC}}^{\mathrm{T}}\mathbf{x} + d_{\mathrm{SOC}}I(t)$. If the state is accurately estimated, then the SOC can be calculated from this equation.

The state estimator equations are

$$\dot{\hat{\mathbf{x}}} = \mathbf{A}\hat{\mathbf{x}} + \mathbf{b}I(t) + \mathbf{l}\mathbf{c}^{\mathrm{T}}\mathbf{e}, \tag{7.30}$$

$$\hat{V}(t) = \mathbf{c}^{\mathrm{T}}\hat{\mathbf{x}} + dI(t), \tag{7.31}$$

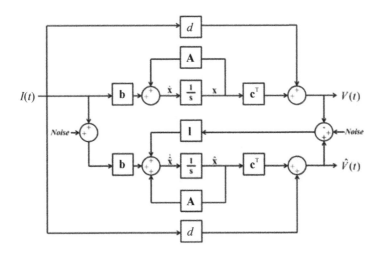

Figure 7.3   State estimator block diagram

where the estimated state $\hat{\mathbf{x}} \in \mathcal{R}^N$, estimated voltage $\hat{V}(t)$, observer gain matrix $\mathbf{l}$, and estimator error $\mathbf{e}(t) = \mathbf{x}(t) - \hat{\mathbf{x}}(t)$ are used. The state estimator can be propogated in real time on board a device if the voltage and current are measured. Integration in Equation (7.30) is typically achieved numerically using a microprocessor with a fixed sample time. If the sample rate is sufficiently fast, then the discrete time approximation implemented on the microprocessor (e.g., Euler integration) will accurately approximate the continuous time integrals.

In practice, the measured current and voltage that are inputs to the estimator are noisy. Also, the system is never perfectly modeled, so there are mismatches between the linear, state-variable model and the actual battery dynamics. The noise and model mismatch, however, are somewhat ameliorated by the feedback term $\mathbf{l}\mathbf{c}^{\mathrm{T}}\mathbf{e}$ in the state equations of the estimator. If the actual output and estimated output differ for any reason (e.g., different initial conditions, noise, and model mismatch) then there is a correction term to the estimate that tends to reduce the error.

An estimator that acts to reduce the error is desired, so the observer gain matrix must be chosen to ensure stable error dynamics. The error dynamics are obtained by subtracting the state equations (7.28) from the state estimator equations (7.30) to produce

$$\dot{\mathbf{e}} = \left(\mathbf{A} - \mathbf{l}\mathbf{c}^{\mathrm{T}}\right)\mathbf{e}. \tag{7.32}$$

These error dynamics are stable as long as the matrix $\mathbf{A} - \mathbf{l}\mathbf{c}^{\mathrm{T}}$ has eigenvalues in the left half of the complex plane. One method to design $\mathbf{l}$ is to place the eigenvalues at specific locations with desired damping and response time using, for example, the Matlab command `place.m`. A variety of other methods of observer design are available in the field of modern control, including techniques designed for stochastic (LQG), uncertain (H^∞), and nonlinear (Kalman filtering) systems.

The state estimator is based on a copy of the cell dynamics with a correction term that feeds back the error between the estimated voltage $\hat{V}(t)$ and the measured voltage. If the estimator and cell have the same dynamics, initial conditions, and input, then the estimated state (and voltage) will perfectly track the actual state. In practice, however, the initial condition is not known, so the estimator uses error information to converge the actual and estimated states. This convergence is guaranteed if the model and the experimental system are identical, there is no sensor noise, and the gain matrix $\mathbf{l}$ is chosen to place the poles of $\mathbf{A} - \mathbf{l}^T\mathbf{c}$ in the left half of the complex plane.

To examine the performance of state estimators in battery systems, two observers are designed for an Ni–MH cell. The first uses the first-order, averaged dynamics described in Section 7.1.1. The second is based on the linearized, single-electrode, third-order model described in Section 6.3.6. Both models are simulated with the second model as the plant, differing initial conditions between the plant and estimator, and voltage and current noise. The plant is linearized at an SOC of 0.7. Random, zero-mean noise of 10 mV and 0.5 A is included in the voltage and current measurements, respectively.

Figure 7.4 shows the results for SOC estimation based on the first-order SOC dynamics model. The initial SOC for the estimator is 60%, which does not match the actual initial SOC of 70%. The estimator pole is placed at -0.10. The estimated voltage converges to the measured value within 30 s. The SOC estimate also moves toward the actual SOC initially

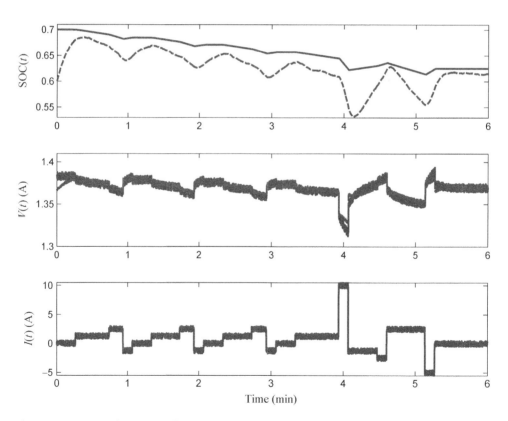

Figure 7.4 SOC estimator for Ni–MH cell based on first-order averaged model: SOC (top), voltage (middle), and current (bottom) with actual (solid) and estimated (dashed) curves

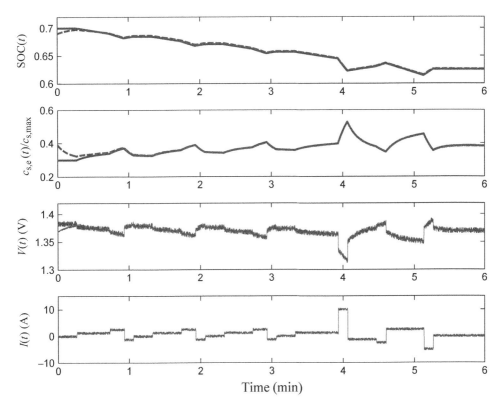

Figure 7.5 SOC estimator for Ni–MH cell based on full-order model: SOC (top), voltage (middle), and current (bottom) with actual (solid) and estimated (dashed) curves

and then starts wandering around as the charge/discharge pulses fire in rapid succession. The mismatch between the SOC model and the cell dynamics and the sensor noise are producing a non-convergent SOC estimate. If the estimator pole is slowed down (e.g., placed at -0.01), then the pulses are more averaged out and the SOC estimate more closely tracks the actual trend. This slow SOC estimator will not follow the peaks and valleys in the SOC curve, however, producing maximum error during large charge/discharge pulses.

Figure 7.5 shows that increasing the accuracy of the underlying model increases the accuracy of model-based estimation. In this case, the same model is used for the plant and for the estimator. The initial conditions on SOC and $c_{s,e}/c_{s,max}$ are 0.69 and 0.40, 1% and 33% different than the actual values. The estimator has three eigenvalues, two of which are placed at the same location as two of the plant poles (-0.013 and -0.112) and the other corresponding to the plant pole at the origin is placed at -0.010. The simulation results show that the estimated voltage converges to close to the actual value in less than 30 s. The surface concentration converges at a slightly slower rate (45 s) and accurately tracks the model value despite the voltage and current noise. SOC converges the least quickly, taking almost 5 min to converge. Once converged, however, it closely tracks the actual SOC even during the current pulses that caused problems for the estimator based on the first-order SOC dynamics model. Moving the estimator poles farther into the left half plane can reduce the convergence time but tends to increase the noise sensitivity, creating more fuzz on the estimated variables. Estimator design involves placing the poles to optimally trade off bandwidth and accuracy. An added advantage of this full-order model-based observer is that internal variables can be estimated. The concentration at the surface of the particles $c_{s,e}(t)$ is a model output, so that variable can be tracked during operation to gain insight into cell operation and avoid operation that may cause damage. Note that this measurement would be difficult if not impossible to obtain in the field.

7.2   Least-Squares Model Tuning

One of the challenges in battery modeling is that there are many parameters that go into the model and the "correct" values are often difficult to determine. The accuracy of a simulation or a model-based estimator is only as good as the underlying model. Battery models have a long list of input parameters, including geometric, electrical, and material properties. It is often difficult, if not impossible, to obtain the model parameters because the manufacturers are reluctant to share the details of cell design and composition. Even battery manufacturers may not know how difficult it is to measure properties like solid phase diffusivity. While there have been many papers specifically dedicated to the measurement of, for example, electrolyte conductivity or diffusivity, the exact composition of the electrolyte and electrodes for a given cell may not be known. The parameters can be set by the adoption of published values, measurement via destructive testing, and measurement of cell voltage response to a variety of current inputs (e.g., see [57]). There are always extra "knobs" to turn (i.e., fitting parameters to adjust) in the model so that the simulated and experimentally measured voltage response match as closely as possible.

The model tuning process has often been based on trial and error but in some cases a least-squares approach can be used. Trial and error depends on the insight and patience of the modeler to adjust the many parameters in some methodical way to get a good match between the simulated and experimentally measured responses. For linear systems one can derive a transfer function and estimate the coefficients of the numerator and denominator polynomials using a least-squares approach. In this way, the tuned model, with parameters identified using a least-square algorithm, is optimal and the resulting coefficients are uniquely determined. In other words, two modelers given the same data will get the same parameters. This provides a firmer foundation for model-based simulation, estimation, and control.

7.2.1   Impedance Transfer Function

Consider the modeled impedance for a battery cell that is expressed in transfer function form:

$$\frac{V(s)}{I(s)} = Z(s) = \frac{b_{n-1}s^{n-1} + \cdots + b_1 s + b_0}{s^n + a_{n-1}s^{n-1} \cdots + a_1 s + a_0}, \tag{7.33}$$

where the numerator and denominator polynomial coefficients are functions of the model parameters. We assume that the order of the numerator is one less than the order of the denominator. This assumption is valid for impedance without a direct feedthrough term (e.g., contact resistance) that makes the order of the numerator equal to the order of the denominator. Identification schemes can be developed for this case as well, and interested readers can find more information in [78, 79].

If we neglect resistance in the Ni–MH model, for example, the impedance transfer function is

$$Z(s) = \frac{b_2 s^2 + b_1 s + b_0}{s^3 + a_2 s^2 + a_1 s}, \tag{7.34}$$

where the numerator coefficients are

$$b_0 = \frac{3840 C^+ [D_s^+]^2}{A F a_s^+ L^+ [R_s^+]^5}, \quad b_1 = \frac{768 C^+ D_s^+}{A F a_s^+ L^+ [R_s^+]^3}, \quad b_2 = \frac{18 C^+}{A F a_s^+ L^+ R_s^+} \tag{7.35}$$

and

$$C^+ = \frac{\partial \eta^+}{\partial c_{s,e}^+} - \frac{2kRT}{F c_{s,max}^+}. \tag{7.36}$$

$\partial \eta^+ / \partial c_{s,e}^+$ is given in Equation (6.101). The denominator coefficients are

$$a_0 = 0, \quad a_1 = \frac{1920 [D_s^+]^2}{[R_s^+]^4}, \quad \text{and} \quad a_2 = \frac{144 D_s^+}{[R_s^+]^2}. \tag{7.37}$$

7.2.2 Least-Squares Algorithm

Figure 7.6 shows a block diagram of the least-squares tuning algorithm. A zero-mean, repeating current sequence is input into the impedance transfer function to produce an output voltage. In

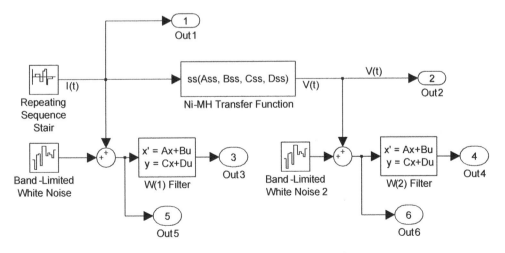

Figure 7.6   Least-squares algorithm block diagram

practice, the voltage and current data would come from experimental testing. Noise is added to both the current and voltage signals to simulate the measurement process. The current and voltage signals are passed through identical filters, represented in state-space form by

$$\dot{\mathbf{w}}_1 = \Lambda \mathbf{w}_1 + \mathbf{b}_\lambda I(t), \tag{7.38}$$

$$\dot{\mathbf{w}}_2 = \Lambda \mathbf{w}_2 + \mathbf{b}_\lambda V(t), \tag{7.39}$$

where

$$\Lambda = \begin{bmatrix} 0 & 1 & \cdots & 0 \\ & & \ddots & \\ -\lambda_0 & -\lambda_1 & \cdots & -\lambda_{n-1} \end{bmatrix} \quad \text{and} \quad \mathbf{b}_\lambda = \begin{bmatrix} 0 \\ \vdots \\ 1 \end{bmatrix}. \tag{7.40}$$

The coefficients $\lambda_1, \ldots, \lambda_n$ are calculated to place the poles of Λ in the left half plane at a desired filtering speed. The filters help reduce the effect of noise and eliminate the need for differentiation of the voltage and current signals, which can only be done approximately and greatly amplifies high-frequency noise. The Laplace transforms of Equations (7.38) and (7.39) have the unique form

$$\frac{\mathbf{W}(s)}{U(s)} = \frac{1}{s^n + \lambda_{n-1}s^{n-1} + \cdots + \lambda_0} \begin{bmatrix} 1 \\ s \\ \vdots \\ s^{n-1} \end{bmatrix}, \tag{7.41}$$

for $\mathbf{W}(s) = \mathbf{W}_1(s)$ with $U(s) = I(s)$ and $\mathbf{W}(s) = \mathbf{W}_2(s)$ with $U(s) = V(s)$. All of the transfer functions in Equation (7.41) are proper (order of the numerator less than or equal to the order of the denominator) and the states $\mathbf{w}(t)$ are zeroth through the $(n-1)$ filtered differentiations of the input signal.

The output voltage can be linearly parameterized in terms of the filter states as follows:

$$V(s) = \mathbf{b}^T \mathbf{W}_1(s) + \mathbf{a}^T \mathbf{W}_2(s) = \Theta^T \mathbf{W}(s) \tag{7.42}$$

where $\mathbf{b}^T = [b_0, \ldots, b_m]$, $\mathbf{a}^T = [a_0 - \lambda_0, \ldots, a_{n-1} - \lambda_{n-1}]$, $\Theta^T = [\mathbf{b}^T, \mathbf{a}^T]$, and $\mathbf{W}^T(s) = [\mathbf{W}_1^T(s), \mathbf{W}_2^T(s)]$.

Linear parameterization can be proven by expanding Equation (7.42) using Equation (7.41) to produce

$$\begin{aligned} \Theta^T \mathbf{W}(s) &= \mathbf{b}^T \mathbf{W}_1(s) + \mathbf{a}^T \mathbf{W}_2(s) \\ &= \frac{b_0 + b_1 s + \cdots + b_{n-1}s^{n-1}}{s^n + \lambda_{n-1}s^{n-1} + \cdots + \lambda_0} I(s) \\ &\quad + \frac{\lambda_0 - a_0 + (\lambda_1 - a_1)s + \cdots + (\lambda_{n-1} - a_{n-1})s^{n-1}}{s^n + \lambda_{n-1}s^{n-1} + \cdots + \lambda_0} V(s) \end{aligned}$$

$$= \left[\frac{b_0 + b_1 s + \cdots + b_{n-1} s^{n-1}}{s^n + \lambda_{n-1} s^{n-1} + \cdots + \lambda_0} \right.$$

$$+ \frac{\lambda_0 - a_0 + (\lambda_1 - a_1)s + \cdots + (\lambda_{n-1} - a_{n-1})s^{n-1}}{s^n + \lambda_{n-1} s^{n-1} + \cdots + \lambda_0}$$

$$\left. \times \frac{b_{n-1} s^{n-1} + \cdots + b_1 s + b_0}{s^n + a_{n-1} s^{n-1} \cdots + a_1 s + a_0} \right] I(s) \tag{7.43}$$

using the impedance transfer function (7.33). Algebraic simplification of the transfer function in the square brackets reduces the right-hand side of Equation (7.43) to the right-hand side of Equation (7.33), so we obtain the equality $\Theta^{\mathrm{T}} \mathbf{W}(s) = V(s)$, proving the linear parametrization in Equation (7.42).

Using the linear parametrization in Equation (7.42), the estimated output

$$\hat{V}(t) = \hat{\Theta}^{\mathrm{T}} \mathbf{w}(t), \tag{7.44}$$

where $\hat{\Theta}$ is the parameter estimate. The error is defined to be

$$e(t) = V(t) - \hat{V}(t) = V(t) - \hat{\Theta}^{\mathrm{T}} \mathbf{w}(t). \tag{7.45}$$

For parameter tuning, we assume that N_{eval} data points (I, V) are provided from system measurements and placed in vectors

$$\mathbf{V}^{\mathrm{T}} = [V(0), V(\Delta t), \ldots, V((N_{\mathrm{eval}} - 1)t)] \tag{7.46}$$

and

$$\mathbf{I}^{\mathrm{T}} = [I(0), I(\Delta t), \ldots, I((N_{\mathrm{eval}} - 1)t)], \tag{7.47}$$

where Δt is the sample time. The sample frequency and N_{eval} must be sufficiently large to capture the high- and low-frequency dynamics of interest, respectively. The current input should be sufficiently active to excite these dynamics as well. These input vectors ($\mathbf{V}$ and $\mathbf{I}$) are fed through the filters to produce

$$\mathbf{J} = [\mathbf{w}(0), \mathbf{w}(\Delta t), \ldots, \mathbf{w}((N_{\mathrm{eval}} - 1)t)]. \tag{7.48}$$

The least-squares cost function

$$CF = |\mathbf{V} - \hat{\Theta}^{\mathrm{T}} \mathbf{J}|^2, \tag{7.49}$$

so the $\hat{\Theta}$ that minimizes the CF is given by

$$\hat{\Theta}_{\mathrm{ls}} = [\mathbf{J}\mathbf{J}^{\mathrm{T}}]^{-1} \mathbf{J}\mathbf{V}. \tag{7.50}$$

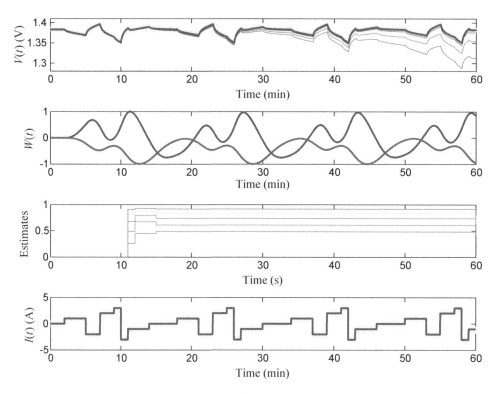

Figure 7.7 Least -squares tuning of a Ni–MH model: (a) measured and estimated voltage with different data sets; (b) normalized filter response; (c) normalized parameter estimates; and (d) actual (thick) and measured (thin) current input

7.2.3 Ni–MH Cell Example

Figure 7.7 shows the simulated response of the simplified Ni–MH model given by the transfer function (7.34). The input current consists of positive and negative current pulses that repeat after 15 min and continue for an hour without changing the SOC. The voltage response changes by ±25 mV in response to the current input. The voltage and current plots also show the small-amplitude noise that has been introduced to simulate realistic sensor signals. The responses of the first states of both the current and voltage filters are shown, normalized with respect to their maximum values so that they can be displayed on the same axis. All three filter poles are placed at −0.01/s, corresponding to a time constant of 100 s. This gives the best trade-off between convergence time and noise sensitivity. The filters clean up the noise quite nicely, limiting the noise impact on the filter output. The other filter states, however, have more noise because they are obtained by time differentiation of the first states.

The parameter estimates are shown for different data sets in Figure 7.7c. The estimates are normalized by their actual values, so if the estimate perfectly matches the actual then the estimates shown in the figure go to unity. Using the first 11 s of data, the parameter estimates jump to their initial values. The parameters have the correct sign and are within a factor of

4 of their actual values. With 12 s of data, the estimates jump towards unity and are only off by a maximum of a factor of 2. Using 15 s of data, the estimates move closer and farther from unity, due to the effects of noise. Using the entire 60 s data record does not move the estimates much from the 15 s data set results. The additional 45 s of data simply repeat the first 15 s, so no additional information can be gleaned by the estimator. In fact, the 60 s estimates are actually farther from the actual values than the 15 s estimates are. The voltage estimates plotted in Figure 7.7a best agree with the actual voltage for the 60 s parameters' estimates shown with the thick line. If the noise is turned off then the parameters converge to their actual values in 15 min. With noise and 10 min of data the parameter estimates result in an unstable system. It is important to use enough data and a sufficiently varied input signal to ensure convergence.

7.2.4   Identifiability

A set of parameters is identifiable for a given system if they can be uniquely determined from the voltage response to a sufficiently rich input current signal. Identifiability depends on both the system and the parameters. For the same system, some parameter sets may be identifiable and others may not. A sufficiently rich input current signal has enough frequency content to fully excite the cell dynamics. As a rule of thumb, one must add another sinusoid at a unique frequency for each parameter to be identified. Typical driving cycles, for example, have no problem providing a rich input current signal.

The previous example provides insight into why systems are often not identifiable. At best, one can identify the coefficients in the numerator and denominator polynomials that make the model response best match the experimental response in a least-squares sense. These coefficients are, in turn, dependent on the model parameters. In the Ni–MH example, there are five coefficients that can be identified. These five coefficients depend on 12 parameters. It is impossible to uniquely determine the physical parameters from the identified coefficients because there are only five equations and 12 unknowns. This is an example of a system that is not identifiable. If seven parameters can be independently determined, then it may be possible to solve for the remaining five parameters using the identified coefficients. The system may still not be identifiable, however. For example, if all of the parameters are known except for A, a_s^+, and L^+, the five equations associated with the coefficients cannot uniquely determine these parameters because they appear in every equation as the product $A a_s^+ L^+$. It may be possible to determine the product but not the individual factors.

In fact, the Ni–MH transfer function depends only on two independent parameters,

$$\alpha_1 = \frac{C^+}{A F a_s^+ L^+ R_s^+} \quad \text{and} \quad \alpha_2 = \frac{D_s^+}{[R_s^+]^2}, \tag{7.51}$$

because Equation (7.34) can be rewritten as

$$Z(s) = \frac{18\alpha_1 s^2 + 768\alpha_1\alpha_2 s + 3840\alpha_1\alpha_2^2}{s^3 + 144\alpha_2 s^2 + 1920\alpha_2^2 s}. \tag{7.52}$$

The least-squares method can estimate the five coefficients in Equation (7.52) that best fit the time-domain data. One can then find the α_1 and α_2 that minimize the error between the estimated coefficients and the model. In the Ni–MH example, using the full 60 s of data, the least-squares transfer function is

$$Z(s) = -\frac{3.2284 \times 10^{-4}s^2 + 7.9321 \times 10^{-6}s + 2.6963 \times 10^{-8}}{s^3 + 0.0802s^2 + 6.3457 \times 10^{-4}s - 2.3808 \times 10^{-8}}. \tag{7.53}$$

Defining a normalized L_2 error norm,

$$e = \sqrt{\left(\frac{18\alpha_1 + 3.2284 \times 10^{-4}}{3.2284 \times 10^{-4}}\right)^2 + \cdots + \left(\frac{1920\alpha_2^2 - 6.3457 \times 10^{-4}}{6.3457 \times 10^{-4}}\right)^2}, \tag{7.54}$$

we can find the the parameters that minimize e using, for example, the Matlab optimization toolbox (fminsearch.m). This yields $\alpha_1 = -1.8833 \times 10^{-5}$ and $\alpha_2 = 5.7551 \times 10^{-4}$, close to their actual values of $\alpha_1 = -2.2616 \times 10^{-5}$ and $\alpha_2 = 8.6957 \times 10^{-4}$.

Alternatively, we can solve for α_1 and α_2 using the identified numerator and denominator coefficients and the known form of the transfer function given in Equation (7.52). In this example, there are five coefficients and only two unknowns. Solving for all combinations of coefficients yields nine distinct solutions for (α_1, α_2). The standard deviation of these solutions provides insight into the accuracy of the estimated model for the given data. The mean of these solutions can be used as a simpler estimate than the minimization technique described previously.

Table 7.1 shows the optimal, solution combinations, and mean parameter estimates for the Ni–MH example. The parameter values are normalized by their actual values so a perfect estimate has a value of 1.0. All of the parameter estimates are lower but within 36% of their actual values. The mean and optimized values are similar so the simpler mean calculation may be warranted for fast computation. The parameter values calculated from the various combinations are consistent, indicating that the model is a reasonable approximation for this system.

Table 7.1 Parameter estimates for Ni–MH example

Method		α_1 Equation	α_1 Value	α_2 Equation	α_2 Value
b_2	b_1	$b_2/18$	0.7930	$3b_1/128b_2$	0.6622
b_2	b_0	$b_2/18$	0.7930	$3b_0/640b_2$	0.7196
b_2	a_2	$b_2/18$	0.7930	$a_2/144$	0.6401
b_2	a_1	$b_2/18$	0.7930	$\sqrt{a_1/1920}$	0.6611
b_1	b_0	$5b_1^2/768b_0$	0.6717	$b_0/5b_1$	0.7818
b_1	a_2	$3b_1/16a_2$	0.8205	$a_2/144$	0.6401
b_1	a_1	$b_1/96\sqrt{30/a_1}$	0.7944	$\sqrt{a_1 1920}$	0.6611
b_0	a_2	$27b_0/5a_2^2$	1.0021	$a_2/144$	0.6401
b_0	a_1	$b_0/2a_1$	0.9394	$\sqrt{a_1/1920}$	0.6611
Mean			0.8076		0.6758
Optimized			0.8327		0.6618

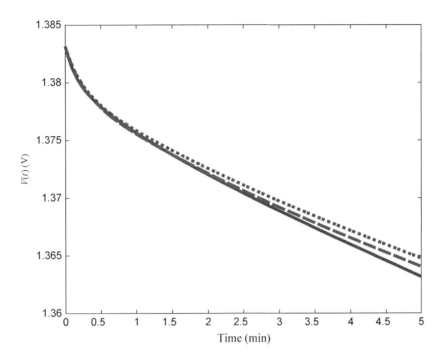

Figure 7.8 Voltage response to unit step discharge current for the Ni–MH model: actual model (solid), optimized (dashed), and mean (dotted)

To better understand the effect of parameter estimation error on performance prediction, the time responses of the mean and optimized models are compared with the actual system. Figure 7.8 shows the voltage response to a unit step discharge current input for the Ni–MH model. The identified models using optimized and mean techniques match well with the actual system response for the 5 min transient shown. The parameter estimates, being smaller than actual, manifest themselves in a steady-state slope offset that produces diverging voltage estimates as time increases.

The Ni–MH example shows us that although the model has 12 input parameters, only two parameters can be uniquely determined from identification of the impedance transfer function. These two parameters are related to the physical parameters through Equations (7.51) and (7.36), but the physical parameters cannot be uniquely determined. For this system, only the composite parameters α_1 and α_2 are identifiable.

7.3 SOH Estimation

The health of a battery is characterized by its capacity and internal impedance or resistance. A healthy battery has close to the manufacturer's advertised capacity in ampere-hours and a "small" impedance. The capacity depends on the volume of active material in the anode and cathode. The capacity of the electrode that the manufacturer is most concerned about degrading is often oversized relative to the other electrode. As the battery is cycled, the electrodes and electrolyte degrade, the capacity decreases, and the impedance increases. There are often many degradation mechanisms acting simultaneously, but a well-designed battery will be balanced so that one degradation mechanism does not cause premature failure.

For the Li-ion chemistry, the capacity given by Equation (7.13) is proportional to the electrode thickness, δ, area A, porosity ε, maximum lithium concentration $c_{s_{max}}$, and stoichiometry $\theta_{100\%}$ and $\theta_{0\%}$. For the capacity to decrease, one or more of these parameters have to decrease. The electrode thickness and area are geometric parameters that do not change. Maximum lithium concentration is a property of the electrode chemistry and is invariant. That leaves porosity and stoichiometry as the parameters most likely responsible for capacity fade. A change in porosity is associated with active mass loss where some of the active material is no longer participating in the reaction. Changes in stoichiometry are associated with the loss of cyclable active material that occurs when it is consumed, for example, in a side reaction.

As we have already seen, however, it may be difficult to identify the parameters that are specifically responsible for capacity loss, so we look for parameters that also change with aging that can be correlated to capacity loss, often using an empirical relationship. Internal impedance rise, for example, has been correlated to capacity fade in Li-ion cells [80]. This correlation can also be motivated by the predominant degradation mechanism, which, if correctly identified, can be used to correlate parameters that may change with aging and are relatively easy to identify with capacity. Thus, the on-line estimation of these parameters can be correlated to SOH even if SOH is not directly estimated. Of course, the degradation mechanisms may also be dependent on use, especially temperature and current rate. The environment and battery usage must be taken into account in a practical SOH estimation algorithm.

The focus of this section is on-line parameter estimation for SOH determination, but several other methods could be used. The least-squares approach described in Section 7.2 could provide an off-line parameter and SOH estimate but is not well suited to SOH estimation in real time on board a device or vehicle. The most accurate and direct way to measure capacity is to do a full discharge at a very slow ($C/10$) rate followed by a very slow full charge. Using current counting, the battery capacity can be determined. This is time consuming and impractical, however, for most applications. A full charge/discharge at a higher C rate can also be correlated with capacity as long as the change in internal impedance is correctly accounted for. Again, for an HEV application, this test is not practical. One can use an empirical approach by counting current cycles, but the required maps are time consuming to obtain, especially considering the variety in pulse charge/discharge shapes and environmental conditions. If a degradation model is available, it can be simulated on board the vehicle to estimate SOH. These models can be computationally demanding, however, and may drift from the actual SOH without feedback from the battery.

7.3.1   Parameterization for Environment and Aging

The performance of a battery depends both on its age and its environment. SOH estimation involves differentiating between the temporary performance change due to extreme temperature and the longer term performance degradation associated with aging. It is important, therefore, to characterize how the model parameters change with temperature and aging. Some parameters may change significantly with age but not change much with temperature and, therefore, would be good indicators of SOH. Other parameters can be assumed constant or their change is not related to aging, so computational power should not be consumed to estimate them. There may be parameters that change with both aging and temperature, but the temperature effect can be calibrated out using the measured temperature and empirical or Arrhenius relationships (see Equation (3.64)).

The change in system parameters due to aging depends on the degradation mechanism in a given cell. If the predominant degradation mechanism can be determined, then the parameters that are most closely associated with that mechanism would be most likely to change. If the degradation mechanism involves unmodeled dynamics in the cell, however, then the correlation between the mechanism and system parameters becomes unclear. Ramadass *et al.* [14] were the first to link cell aging to the change of only a few parameters in an electrochemical battery model. For an Li-ion cell, they found that the solid electrolyte film resistance and the solid-state diffusion coefficient of the anodic active material are linked to cell degradation. Goebel *et al.* [80] related the capacity of an Li-ion cell to the easily estimated internal impedance, showing a linear dependency. Schmidt *et al.* [70] found that electrolyte conductivity and cathodic porosity are key parameters to estimate the rate capability fade and the capacity loss of an Li-ion cell. Similar studies can be found for Ni–MH and Pb–acid chemistries.

7.3.2 Parameter Estimation

Parameter estimation for linear systems is a well-established field with several excellent textbooks for interested readers [29, 78, 79]. In this section, a gradient parameter estimator is introduced for continuous-time, SISO systems. More advanced techniques exist to systematically handle the effects of noise that use more advanced estimation laws (e.g., least squares), use prior information (known parameters) to reduce the estimator order, and include nonlinearity.

Figure 7.9 shows a block diagram of a gradient-based parameter estimator. The plant transfer function has current input and voltage output as given by Equation (7.33). The objective is to estimate the parameter vector $\Theta^{\mathrm{T}} = [\mathbf{b}^{\mathrm{T}}, \mathbf{a}^{\mathrm{T}}]$ from the voltage and current data

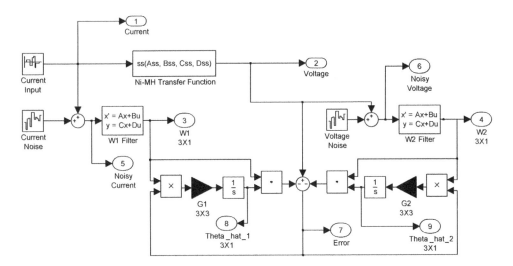

Figure 7.9   Gradient-based parameter estimator for SISO linear systems

in real time using a recursive algorithm that continually updates the parameter estimates as information becomes available. The parameter estimator consists of input $\mathbf{W}_1(s)$ and output $\mathbf{W}_2(s)$ filters given by Equation (7.41) and two gradient estimators that produce estimates of the numerator $\mathbf{b}$ and denominator $\mathbf{a}$ coefficients. The gradient update laws

$$\dot{\hat{\Theta}}_1 = e(t)\mathbf{G}_1\mathbf{w}_1(t) \tag{7.55}$$

$$\dot{\hat{\Theta}}_2 = e(t)\mathbf{G}_2\mathbf{w}_1(t) \tag{7.56}$$

are dynamic equations that are integrated in real time to produce the time-varying estimates $\hat{\Theta}_1(t)$ and $\hat{\Theta}_2(t)$. The gradient update laws depend on the filter current and voltage filter outputs and the error

$$e(t) = V(t) - (\hat{\Theta}_1^{\mathrm{T}}(t)\mathbf{w}_1(t) + \hat{\Theta}_2^{\mathrm{T}}(t)\mathbf{w}_2(t)) \tag{7.57}$$

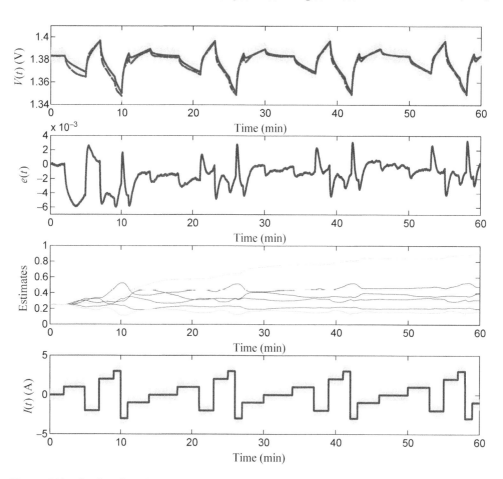

Figure 7.10 Gradient-based parameter estimator simulation for Ni–MH cell example: actual (solid), estimated (dashed), and noisy (light) voltage, voltage estimate error, parameter estimates, and input (solid) and noisy (light) current

between the estimated voltage $\hat{V}(t) = \hat{\Theta}_1^T(t)\mathbf{w}_1(t) + \hat{\Theta}_2^T(t)\mathbf{w}_2(t)$ and the measured voltage. In the absence of noise, one can show that the parameter estimates are bounded. With persistently exciting voltage inputs, the parameter estimates may converge to their actual values.

7.3.3   Ni–MH Cell Example

For the Ni–MH example we use the same parameters and filters as in Section 7.2.3. The gradient update gain matrices are diagonal with diag$\{\mathbf{G}_1\} = \{4 \times 10^{-14}, 8 \times 10^{-10}, 1.6 \times 10^{-6}\}$ and diag$\{\mathbf{G}_2\} = \{4 \times 10^{-11}, 4 \times 10^{-6}, 4 \times 10^{-2}\}$ obtained by trial and error. The parameter estimate integrators are initialized to 25% of their actual values. The time response shown in Figure 7.10 includes the voltage, error, estimates, and current input. The actual voltage and current are contaminated with noise and these noisy signal are filtered to produce $\mathbf{w}_1(t)$ and $\mathbf{w}_2(t)$, respectively. The voltage error is calculated based on the filter outputs and the parameter estimates according to Equation (7.57). The noise prevents the error signal from converging to zero, mainly because the estimates are not converging to their actual values. Some of the estimates, in fact, appear to be heading in the wrong direction. The error can be made to converge to zero if the noise is reduced, the estimates are initialized closer to their actual values, a longer data record with more and varied pulses is used, and the adaptation

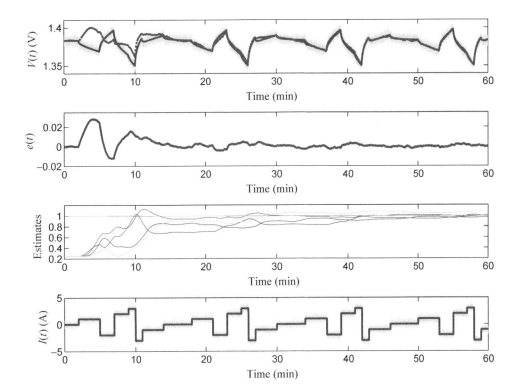

Figure 7.11 Gradient-based parameter estimator simulation for Ni–MH cell example: actual (solid), estimated (dashed), and noisy (light) voltage, Voltage estimate error, parameter estimates, and input (solid) and noisy (light) current

gains are adjusted. For example, if two of the denominator coefficients are initialized at their actual values and their gains are reduced, then the result in Figure 7.11 is obtained. In this case, the parameter estimates converge to their actual values and the error converges toward zero despite the noisy voltage and current signals.

Thus, parameter estimation can be used to estimate the coefficients of a battery cell transfer function if the input current is sufficiently rich, the parameters are initialized close to their actual values, the noise is low, and the gains are correctly chosen. These coefficients can then be correlated to the model parameters and used to track the slow evolution of the battery SOH.

Problems

7.1 An impedance transfer function for a 6 Ah Li-ion cell is

$$\frac{V(s)}{I(s)} = \frac{-0.001\,93s^3 - 0.000\,268\,2s^2 - 3.873 \times 10^6 s - 9.026 \times 10^9}{s^3 + 0.069\,93s^2 + 0.000\,474\,4s}. \quad (A.7.1)$$

This transfer function is linearized at an open-circuit potential of 3.6 V and 50% SOC.

(a) Convert this transfer function to a third-order state-space model with one of the states equal to the $SOC(t)$. Simulate the response to the duty cycle input in Figure 6.11. Plot the response in a single figure with the top subplot $SOC(t)$, $SOC_{cc}(t)$, and $SOC_{vl}(t)$, middle subplot $V(t)$, and bottom subplot $I(t)$, where $SOC_{cc}(t)$ is the SOC estimate using the current counting method and $SOC_{vl}(t)$ is the SOC estimate using the voltage lookup method. Assume that there is an initial SOC estimate error: $SOC_{cc}(0) = 0.9SOC(0)$ and that the voltage sensor has 10 mV (amplitude) of random noise. Discuss the performance of these two methods.

(b) Design an SOC estimator based on a first-order model that only includes the SOC state. Adjust the estimator pole placement to get the best performance. Assume that there is an initial SOC estimate error: $SOC_{est}(0) = 0.9SOC(0)$ and the voltage sensor has 10 mV (amplitude) of random noise. Simulate the response to the duty cycle input in Figure 6.11 and plot the response using the same layout as in part (a). Discuss the performance of this method.

(c) Design an SOC estimator based on the full-order model. Adjust the estimator pole placement to get the best result. Assume that there is an initial SOC estimate error: $SOC_{est}(0) = 0.9SOC(0)$, the rest of the initial state estimate is zero, and the voltage sensor has 10 mV (amplitude) of random noise. Simulate the response to the duty cycle input in Figure 6.11 and plot the response using the same layout as in part (a). Discuss the performance of this method.

7.2 A state-space model of an 8.75 Ah Pb–acid cell, linearized at 70% SOC (2.12 V), has the state matrices

$$\mathbf{A} = \begin{bmatrix} 0 & 0 & 0 \\ 0.5113 & -0.010\,05 & 0 \\ 0 & 0 & -0.2727 \end{bmatrix}, \quad \mathbf{b} = \begin{bmatrix} 1.332 \times 10^{-6} \\ 0.000\,290\,7 \\ -0.002\,907 \end{bmatrix}, \quad (A.7.2)$$

$$\mathbf{c}^T = \begin{bmatrix} 0 & 1 & -1 \\ 23.83 & 0 & 0 \end{bmatrix}, \quad \mathbf{d} = \begin{bmatrix} 0 \\ 0 \end{bmatrix}, \quad (A.7.3)$$

where the first output is the voltage $V(t)$ and the input is the current $I(t)$.

(a) Show that the second output is SOC.

(b) Simulate the response to a square wave input with 1 A amplitude and a 100 s period for six complete cycles. Plot the response of the voltage, SOC, and current.

(c) Calculate the gain matrix for a full state estimator that places the poles at $[-0.01, -0.1, -0.27]$.

(d) Simulate the estimator response to the same input as part (b) with zero initial conditions on the estimated state and estimated SOC. Plot the response of the voltage and voltage estimate (same plot), SOC and SOC estimate (same plot), and current. Produce a second plot with 10 mV of voltage sensing noise and 10 mA of current sensing noise.

(e) Design a least-squares tuning algorithm for this model that uses a batch of data from one cycle to calculate the transfer function coefficients. Simulate the batch least-squares tuning algorithm using the input in part (b) and plot the results using the format in Figure 7.7. Adjust the filter poles to give the best results. Produce a second plot with 10 mV of voltage sensing noise and 10 mA of current sensing noise.

(f) Design a parameter estimator for this model that calculates the transfer function coefficients in real time. Simulate the parameter estimator using the input in part (b) and plot the results using the format in Figure 7.11. Use the filter poles from part (e) and adjust the estimation gains to get the best results. Start the simulation with initial estimates at 25% of their actual values. Produce a second plot with 10 mV of voltage sensing noise and 10 mA of current sensing noise.

8

Battery Management Systems

A BMS is responsible for controlling the charge and discharge currents going into and out of the battery pack, respectively. The primary responsibilities of the pack BMS are (i) limiting overcharge and undercharge in the cells, (ii) ensuring that the cells in the pack are balanced, and (iii) maintaining safe operation of the pack. The large number of cells, wiring harness, and electromechanical complexity of a typical HEV battery pack lead to significant challenges for battery pack design and management.

For many applications, the battery is slowly discharged over time (e.g., laptop battery) and then recharged periodically. In this case, the BMS is primarily concerned with recharging the battery in a short time to full charge with a charging protocol that does not damage the battery and reduce its life. Of course, if the device is plugged in then the BMS can either maintain a full charge on the battery pack and draw outlet power, run off the battery without drawing outlet power, or some combination of the two. For some battery chemistries, longer battery life can be achieved by allowing the battery to discharge periodically and not keeping it fully charged when the device is plugged in. The user expects a fully charged battery for maximum usage time, however, so life extension must be balanced with the desired use.

The most complex BMSs are required to deal with large charge and discharge currents in a dynamic environment. In HEVs, for example, the batteries are continually being charged and discharged based on the supply and demand of motive power. At every instant in time, the BMS must decide whether to direct all or part of the charge/discharge current into/out of the battery pack. Charge current resulting from regenerative braking can either be sent to the pack or dissipated in a resistor bank. If current or voltage limits will be exceeded, then mechanical braking can be engaged. Dissipation of energy in resistors or mechanical brakes, however, reduces the overall efficiency of the HEV. The demanded discharge current can similarly be limited to prevent undervoltage or overcurrent operation.

If limits on current and voltage are fixed in hardware, then a higher level, supervisory controller typically makes decisions to avoid these limits. The decisions and implementation of the decisions take time, however, so it is useful for the supervisory controller to include performance predictions in the decision process, such as the maximum current that can be supplied/withdrawn from the battery over the next Δt seconds. Given the cell states, SOC, SOH, and temperature, the maximum charge and discharge current limits are predicted such

Battery Systems Engineering, First Edition. Christopher D. Rahn and Chao-Yang Wang.
© 2013 John Wiley & Sons, Ltd. Published 2013 by John Wiley & Sons, Ltd.

that over- and under-voltage conditions will not occur in the next Δt seconds, assuming constant current rate. The time period Δt can be chosen to provide feedforward information to alternative power source/sinks such that they can be engaged to fill the gap between demanded and available power from the battery pack.

During operation, it is important that the individual cells that make up the pack are balanced. A perfectly balanced pack has all cells at the same SOC and, hence, voltage. The cells in a pack are connected in parallel to increase the available current and in series to increase the operating voltage. Cells connected in parallel are automatically balanced because they have the same output voltage. Current will flow through the parallel cells inversely proportional to their impedance, however. Cells connected in series have the same current. Theoretically, if the cells in a series string are identical and subject to the same environment, they would remain balanced if they start out balanced. Unfortunately, even fresh cells are not identical and each cell ages in a unique way. The performance and aging also depend on the environment. Packs can be quite large and cells on the outside see a different temperature history than cells on the inside. The performance and aging of cells is highly dependent on temperature; so, even if the cells start out identical, they will perform and age differently over time. Finally, even tiny parasitic loads from, for example, the BMS itself, if not uniformly applied to all the cells, will unbalance a series string.

The selection and heating/cooling of cells in a pack can minimize cell-to-cell differences. Cells are often capacity tested prior to assembling the pack to ensure that cells with closely matched capacity are grouped in a series string. Heating and cooling channels can be put in the pack to minimize thermal gradients. These steps, however, cannot entirely eliminate cell differences that lead to pack unbalance.

Differences in the string are often amplified by degradation. Consider water loss in a Pb–acid cell, for example. Water loss results from overcharging the cell and reduces the cell's capacity. During charge, a cell with water loss will reach full capacity sooner than the other cells in the pack. If the charging continues to bring the other cells to 100% SOC, then the degraded cell will experience overcharge, lose more water, and reduce capacity even further. With cell-level control of charging, the BMS would know to stop charging the degraded cell before the other cells in the string.

Pack safety is another important responsibility of the BMS. Keeping the cell currents and voltages within safe limits provides one level of safety. Monitoring the cell temperatures provides another level of safety. Thermal management systems maintain the pack within a specified temperature range. Heat is generated in the pack by the electrochemical reactions, Joule heating (I^2R), and other reversible and irreversible electrode processes. Thermal management is critical because electrochemical reaction rates increase exponentially with temperature, often doubling with every 10 °C of temperature rise, but heat dissipation rates increase only linearly with temperature. As a result, catastrophic thermal runaway may occur with the potential for an explosive event. Lithium batteries, in particular, have 20% of the energy density of common explosives, so the potential for a dangerous event is always possible. Uniformity of temperature is also important for maintaining the balance between cells in series strings. Temperature gradients in the pack result in nonuniform degradation amongst the cells, complicating cell balance and reducing pack efficiency. The battery systems engineer should also consider the effects of pack physical damage from, for example, physical penetration that shorts two or more current collectors. Active and passive (i.e., fused) pack switching during these extreme events may be warranted.

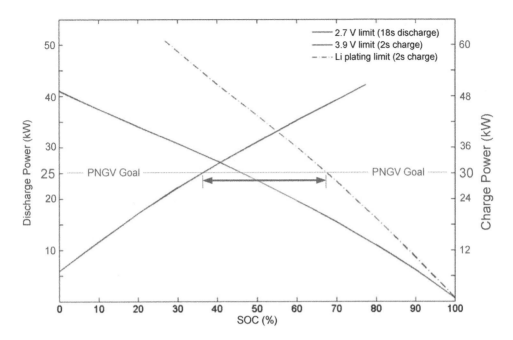

Figure 8.1 Power versus SOC for Li-ion pack: 18 s discharge pulse to a 2.7 V voltage limit (left line) and 2 s charge pulse to a 3.9 V limit (right line) or lithium plating limit (right dash–dotted line)

The development and implementation of a BMS can have different levels of sophistication. Charge/discharge current and voltage limits can be fixed in hardware at conservative values that are safe regardless of operating environment, temperature, and usage history. These limits can be determined empirically based on extensive testing in environmental chambers. Lookup tables can be developed that adjust the current and voltage limits based on the battery age and temperature. At the other end of the spectrum, model-based approaches that take into account the actual processes inside the cell are the most sophisticated, requiring extensive modeling, validation, and analysis. Most applications demand the high performance, maximal pack utilization, and reduced pack size that results from this approach.

Figure 8.1 shows the potential benefit of using dynamic, model-based charge limits for a Li-ion pack. During charge, two different current-limiting strategies are shown. First, a fixed limit based on pack voltage (left) and second a dynamic limit based on a model-based estimator (right). The dynamic limit is adjusted based on the estimated minimum solid–electrolyte potential and is designed to avoid lithium plating, a predominant degradation mechanism involving a side reaction that occurs at low solid–electrolyte potential. Using a dynamic limit can broaden the SOC operating range, as shown by the dash–dotted curve in Figure 8.1. With dynamic current limits, the upper SOC that can provide a 30 kW charge pulse for 2 s increases from less than 50% to almost 70%. Thus, a smaller pack with dynamic current limits can provide the same amount of energy as a larger pack with static current limits. In this example, dynamic current limits provide three times the usable energy and a 22% increase in usable power over conservative, fixed current limits.

In this chapter we apply the models and estimators developed in the previous chapters to the development of an efficient and effective BMS. Charge protocols, available charge/discharge power estimating, charge/discharge limiting, and cell balancing are studied.

8.1 BMS Hardware

Commercially available BMS hardware includes high-power switching elements, cell-balancing integrated circuits (ICs or chips), and current, voltage, and temperature sensors. Figure 8.2 shows an example BMS from Texas Instruments (TI). The BMS supervises and

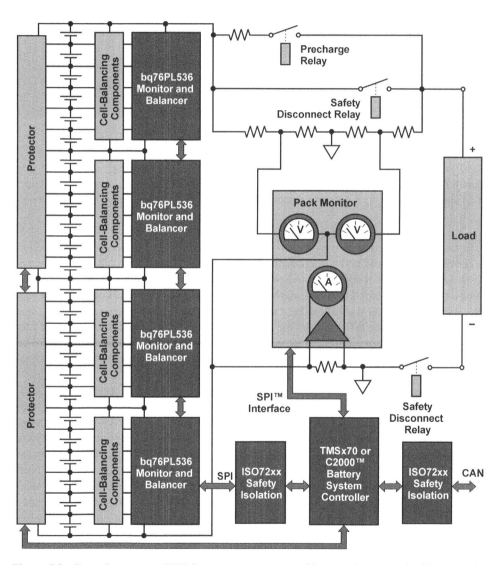

Figure 8.2 Texas Instruments HEV battery management architecture (reproduced with permission from Texas Instruments)

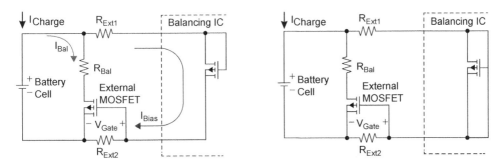

Figure 8.3 Principle of passive cell balancing [81]: (left) external MOSFET "on" with current bypassing cell and (right) external MOSFET "off" with current charging cell (reproduced with permission from Texas Instruments)

cell-balances the battery pack, using different communication paths to ensure system redundancy. Built-in temperature management is included to ensure safe and long life operation. The BMS is a highly safety-critical part of the HEV, so reliable communication and accurate data measurement are necessities. As with any commercial hardware, isolation circuits are provided to protect the sensitive microprocessors (e.g., battery system controller) from the high-power pack voltages.

The TI bq76PL536 chip shown in Figure 8.2 can be used for passive balancing using an external shunt resistor and either the internal (low current) field-effect transistor (FET) or an external, high-power MOSFET metal oxide (MOSFET) [81]. TI also has ICs that do active balancing using capacitive or inductive charge storage and shuttling to move energy between adjacent cells in a pack. Active balancing is more efficient because energy is not dissipated in external resistors, but it uses additional components and is more costly.

The principle of passive balancing using the bq76PL536 is shown in Figure 8.3. Passive balancing can only be performed during charging with the current $I_{Charge} > 0$. If a pin on the IC is activated, then current I_{Bias} flows through an internal FET, turning on an external MOSFET. The external MOSFET then allows current I_{Bal} to bypass the battery cell, dissipating energy in resistor R_{Bal}. If the pin is disabled, the internal FET is turned off and all the I_{Charge} current flows through the cell. In this case a high-power MOSFET is used to allow high current I_{Bal} and fast balancing. Slower balancing can be achieved using only the internal FET and the external resistors R_{Ext}. Each bq76PL536 chip can handle up to six cells and can be stacked vertically for up to 192 cells.

The simplest algorithm for cell balancing is based on measured cell voltage. Cells in a pack have differing impedance, however, so voltage-based balancing may not equalize SOC in a series string. With the nonzero current flow required in passive balancing, differing impedance offsets the cell voltages from each other even if their SOCs are identical. This can be minimized if balancing is only done when the current is small. Limiting the balancing currents, however, increases the balancing time. The balancing currents must be at least as large as the pack imbalance or the cell SOCs will slowly drift apart.

Cell balancing can also be based on charge, using, for example, a TI bq20z80 Impedance Track™ gas gauge chip [82]. The chip uses voltage lookup to determine SOC after the battery has been resting for a sufficiently long time at current levels less than $C/20$. Voltage versus SOC and temperature are provided for various chemistries. Current counting is then employed to estimate SOC. The cell capacity is updated based on two current counting DOD estimates in a relaxed state. The relaxed state is determined by waiting until dV/dt crosses a threshold or elapsed time without charge/discharge reaching a limit, typically several hours. Cell impedance (high-frequency or ohmic resistance) is calculated during discharge by calculating the difference between the open-circuit voltage and measured voltage and dividing by discharge current. With estimates of capacity and SOC for each cell in the pack, charge can be moved around the pack to balance SOC.

The workhorses of the BMS are high-power FET switches. Signal FETs can handle low (5 V) voltages. Power MOSFETs can cover drain-to-source voltages up to 200 V. Above 200 V, insulated-gate bipolar transistors (IGBTs) are typically used. IGBTs are hybrids of MOSFETs and bipolar junction transistors that are heavily used in HEV drive electronics.

8.2 Charging Protocols

The objectives of charging are to restore the battery to full state of charge as efficiently as possible and in as short a time as possible. Charging protocols typically involve constant voltage (CV), constant current (CC), or combinations of constant voltage and constant current (CC–CV) charging. Under CV charging, the current rushes in quickly for a low SOC battery and then slowly tapers off as the battery voltage rises. If the CV level is set appropriately, the current tapers off to almost zero as the battery SOC reaches 100%. CC charging is typically applied as a starting charge method. The CC–CV method starts with a constant applied current until the battery voltage reaches a cut-off value and then the charger switches to CV charging. One can only control the current flowing into the battery or the charger voltage (not both) at any given time. There are many different ways of determining appropriate cut-off voltages, using timers, and switching between CC and CV charging [2].

Charging schemes are often applied to many different batteries at different temperatures and ages, so it is important to consider the variability between different batteries and how that effects the charging process. The CV level, for example, may have to be adjusted based on the performance of the battery being charged. In a large battery pack this becomes even more complicated because of cell variability.

Charging is when the battery is most prone to catastrophic failure. A smart charger must be able to identify when a battery is heating up and may fail due to a condition called thermal runaway. Batteries are high-energy-density devices and can explode under certain conditions. The charging algorithm must have safe charging as the number one priority.

Figure 8.4 shows an example charge profile for an Li-ion battery. If the battery voltage is below $V_{O(LOWV)}$ at the start of charging, the battery is assumed to be discharged and a pre-charge current is applied at $I_{O(PRECHG)}$ until the voltage reaches $V_{O(LOWV)}$. Then the charger switches to thermal regulation phase where the current jumps up until the cell or charger temperature reaches $T_{(THREG)}$. If the temperature remains below $T_{(THREG)}$ then the fast-charge current $I_{O(OUT)}$ is applied until voltage reaches $V_{O(REG)}$. Then current is slowly reduced to maintain the cell voltage at $V_{O(REG)}$. When the current reaches $I_{(TERM)}$ the charge is complete and the charger is turned off.

PHEVs and EVs need dedicated chargers for fast, convenient, and affordable recharging of the vehicle's batteries. Owners of PHEVs and EVs will need residential chargers for their vehicles. A rich charging infrastructure enables smaller on-board battery packs and lower cost

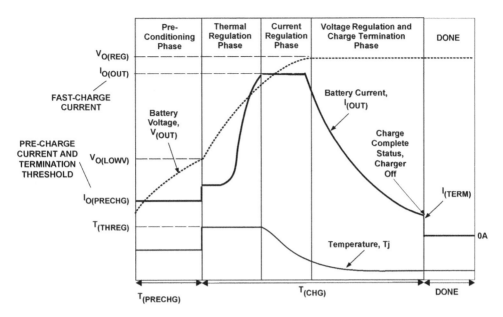

Figure 8.4 Charge profile for an Li-ion battery (reproduced with permission from Texas Instruments)

for PHEVs and EVs. If drivers have many options for recharging, they may select a lower cost vehicle with a smaller battery pack and accept the smaller driving range. Smaller battery packs also charge more quickly; so, if a sufficiently powerful charger is available, the user can recharge during a short shopping trip.

Charging stations typically use a CC–CV charge algorithm, where the charge current tapers off at a CV to a cut-off current. The cutoff current that provides 100% SOC varies based on battery chemistry and manufacturer. Some manufacturers recommend that charging stop after 30 min even if the cut-off current has not been reached. Standards for Level 1, 2, and 3 chargers have been developed that use 120 VAC, 240 VAC, and 480 VAC power, respectively.

Table 8.1 shows the pack sizes and charge times for batteries that provide electric-only driving ranges of 10, 20, and 40 miles. As the size of vehicle increases, the battery size increases for the same range. Level 1 charging times are more than five times longer than Level 2 charging times. A fast charge to roughly 80% SOC can be achieved in shorter times because it eliminates the CV finishing charge. For comparison, a Level 3 charger can charge a dead battery to 50% SOC in 10–15 min, similar to a commercial gasoline service station.

8.3 Pulse Power Capability

In vehicle applications, there are multiple levels of powertrain control. The focus of this book is cell-, battery-, and pack-level estimation and control. Above this low-level control, a supervisory controller determines the power flow between the battery pack, ICE (if used), and electric machines that provide motive force and regenerative braking. The supervisory

Table 8.1 Battery sizes and charge times for EVs [83]

Electric-only range (miles)	10	20	40
Battery size (kWh)			
Economy vehicle	5	10	20
Mid-size vehicle	6.7	13.3	26.7
Light-duty truck/SUV	8.3	16.7	33.3
Approximate charge times – Level 1 (h)			
Economy vehicle	2.7	5.5	10.9
Mid-size vehicle	3.6	7.3	14.5
Light-duty truck/SUV	4.5	9.1	18.2
Approximate charge times – Level 2 (h)			
Economy vehicle	0.5	1.0	2.0
Mid-size vehicle	0.67	1.3	2.67
Light duty truck/SUV	0.83	1.67	3.33

controller can benefit from preview information about the performance limits of the battery pack in the near future. For example, estimation of power available from the pack over the next Δt seconds is valuable to prepare the other power sources/sinks to provide/dissipate any power that cannot be absorbed by the battery pack. Preview windows of $\Delta t = 5$ s or 10 s are typical.

Prediction of maximum current available for Δt seconds can be used in coordination with vehicle supervisory controllers to enable efficient and reliable HEV powertrain control. Depending on computational power, embedded controllers may implement these algorithms either on a single cell or a complete battery pack basis. The latter would require a more conservative approach when setting constraint limits due to cell-to-cell variation. Given the safety issues associated with overcharging, it is already standard practice to measure the voltage of every cell in an Li-ion battery pack.

Battery power is an inherently nonlinear quantity, but it can be linearized at a setpoint to produce the approximation

$$P(t) = I(t)V(t) \approx \bar{P} + \bar{V}\tilde{I}(t) + \bar{I}\tilde{V}(t), \tag{8.1}$$

where $\bar{P}$, $\bar{V}$, and $\bar{I}$ are the setpoint power, voltage, and current with $\bar{P} = \bar{I}\bar{V}$. The deviations from the setpoint are $\tilde{V} = V - \bar{V}$ and $\tilde{I} = I - \bar{I}$. The voltage deviation can be calculated from

$$\tilde{V}(s) = Z(s)\tilde{I}(s), \tag{8.2}$$

where the battery impedance transfer function $Z(s)$ is also linearized at $\bar{V}$ and $\bar{I}$. If the voltage is fairly constant, then current and power are simply proportional to each other during operation, and one can neglect the second term on the right hand side of Equation (8.1). For precise power prediction, however, the full nonlinear Equation (8.1) must be used. Even an LiFePO$_4$ cell, which has a relatively flat voltage versus SOC curve, can have a usable voltage range of 1.6–3.8 V, a range of over a factor of two.

During charge and discharge, the voltage rises and falls, respectively. The longer the pulse, the more the voltage deviates from its initial value. To maintain a constant power, the current must fall and rise during charge and discharge, respectively. The pulse power limit is attained when a variable crosses a predefined limit. In a typical pack, only the voltage, current, and temperature are measured. Although absolute temperature limits are hard wired into the pack to prevent thermal runaway, the temperature responds too slowly to be used as a pulse charge/discharge limit. The simplest approach is to give voltage and current fixed limits,

$$V_{min} \leq V(t) \leq V_{max}, \tag{8.3}$$

$$I_{min} \leq I(t) \leq I_{max}, \tag{8.4}$$

where the lower (V_{min}, I_{min}) and upper (V_{max}, I_{max}) limits are functions of temperature and age. More complicated approaches could include limits on internal variables that, if exceeded, may cause battery degradation [67].

Pulse power capability depends on the battery usage history over both the short term (minutes) and long term (months) and the current environment. In order to predict the pulse power capability, the current state must be known. For a simple first-order SOC model, for example, the current SOC must be known. More complicated models have more states that may include the electrolyte and solid-phase transport dynamics and SOC. In these models, the entire state should be known. In practice, only the voltage is measured, so the state (including SOC) must be estimated. Pulse power capability also depends on the current SOH of the battery. Capacity and model parameters must be identified so that the response can be accurately predicted. Finally, the model parameters also depend on temperature, so their values should be adjusted based on the measured temperature.

If the model parameters and initial state are known, the pulse power capability can be calculated. First, an appropriate end of pulse indicator should be chosen. Voltage limits for charge and discharge are a natural choice. Other internal variables could also be estimated and used. In practice, multiple limits on voltage, current, temperature, and so on will be in force and the BMS must enforce them all. For demonstration purposes, assume that a battery model has been created that predicts voltage given current and is represented in state-space form by

$$\dot{x}(t) = Ax(t) + bP(t) \tag{8.5}$$

$$V(t) = c^{T}x(t) + dP(t), \tag{8.6}$$

where the input is power $P(t)$ and the second term on the right hand side of Equation (8.1) has been neglected.

To predict pulse power capability over the next Δt seconds, the continuous time model in Equations (8.5) and (8.6) is discretized in time with a step size of Δt seconds. Using a zero-order hold (ZOH), the discretized model gives exactly the same solution as the continuous time model at $t = \Delta t, 2\Delta t, \ldots$ if the current pulse is constant during each sample period and changes only at multiples of Δt. To predict the maximum, constant power pulse that can be provided over the next Δt, we can use the ZOH discretized equations.[1]

---

[1] If a continuous time estimator or controller is implemented on a microprocessor, the governing equations must be discretized in this way to produce difference equations that can easily be converted to computer code.

The discretized state equations are

$$x(k+1) = \mathbf{F}x(k) + \mathbf{g}P(k) \tag{8.7}$$

$$V(k) = \mathbf{c}^{\mathrm{T}}x(k) + dP(k), \tag{8.8}$$

where k is the time step ($t = k\Delta t$),

$$\mathbf{F} = e^{\mathbf{A}\Delta t} \quad \text{and} \quad \mathbf{g} = \int_0^{\Delta t} e^{\mathbf{A}\tau} \mathbf{b}\, d\tau. \tag{8.9}$$

The Matlab command c2d.m can also be used to generate the discretized state matrices.

During a constant power charge starting from an initial condition $x(0)$, the voltage reaches the upper limit $V_{\max}$ in Δt seconds if

$$V_{\max} = \mathbf{c}^{\mathrm{T}}x(1) + dP_{\max} = \mathbf{c}^{\mathrm{T}}\{\mathbf{F}x(0) + \mathbf{g}P_{\max}\} + dP_{\max}, \tag{8.10}$$

using Equation (8.7). Solving Equation (8.10), we obtain the charge pulse power capability

$$P_{\min} = \frac{V_{\max} - \mathbf{c}^{\mathrm{T}}\mathbf{F}x(0)}{\mathbf{c}^{\mathrm{T}}\mathbf{g} + d}. \tag{8.11}$$

During a constant power discharge, the pulse power capability is

$$P_{\max} = \frac{V_{\min} - \mathbf{c}^{\mathrm{T}}\mathbf{F}x(0)}{\mathbf{c}^{\mathrm{T}}\mathbf{g} + d}. \tag{8.12}$$

If the current state, voltage limit, and discretized state matrices are known, then the charge and discharge pulse power capability can be calculated from Equations (8.11) and (8.12). In practice, however, the full state is not known, so the state estimate $\hat{x}(0)$ must be used in Equations (8.11) and (8.12). The state estimator also provides a filtering effect on the noisy voltage and current data. Other than during startup, when the initial state is unknown, the state estimate and the actual state should match fairly well.

Figure 8.5 shows an Ni–MH model simulation with pulse power capability estimation from Equations (8.11) and (8.12). The voltage responds to the aggressive power profile with large swings. Arbitrary limits at 1.4 V and 1.3 V are shown being exceeded at $t = 242$ s (discharge) and almost exceeded at $t = 359$ s (charge). The current input is calculated from a desired power profile shown as the dashed line. For most of the simulation the desired power lies under the actual power (light solid) curve. During the deep discharge at $t = 242$ s, however, the actual power is visibly lower than desired due to voltage droop. The dashed curves above zero represent the discharge pulse power capacity and the dashed curves below zero are for charge. The light curves are lower in magnitude because they correspond to longer pulses ($\Delta t = 20$ s) compared with the darker curves ($\Delta t = 5$ s). The curves predict the voltage crossing 1.3 V at $t = 242$ s and touching 1.4 V at $t = 359$ s. The pulse power capability can provide a real-time estimate to the supervisory controller to enable efficient, safe, and optimal switching between various power sources.

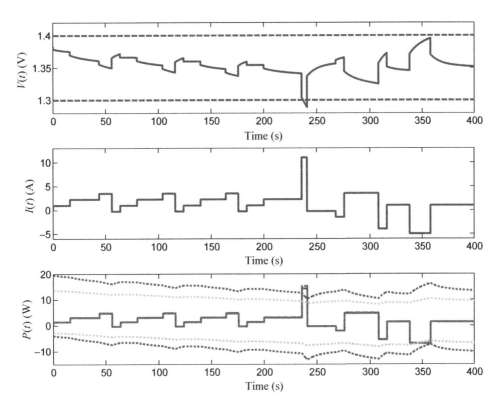

Figure 8.5 Pulse power capability for an Ni–MH cell: voltage response (solid) and limits (dashed), current profile, and power (light solid) and linearized estimate (dashed) and pulse power capacity (5 s (dashed) and 20 s (light dashed))

8.4 Dynamic Power Limits

The power supplied by the pack during discharge and absorbed by the pack during charge is limited. Large currents cause the cell to heat up, possibly resulting in dangerous thermal runaway and/or life-shortening degradation. Large currents can cause the voltage to rise or drop precipitously due to cell impedance. High overvoltage and low undervoltage have the potential to damage the cell. In pulsed power applications, fixed power limits can be overly conservative, particularly for short-duration, high-rate power pulses that give rise to large ohmic voltage perturbations [84]. Manufacturers sometimes rate high-power batteries with multiple power/voltage limits that depend on the duration of the pulse event. These control limits are difficult to realize in a practical BMS because they require future information about how long a pulse will last.

BMS hardware allows programmable power limits for packs and, with cell balancing, individual cells. The simplest method is to pick conservative power limits based on worst-case operating conditions. These limits could then be adjusted based on the temperature, SOH, and SOC of the cell. This requires a three-dimensional map of the operating space and accurate temperature measurement and SOC and SOH estimation. A more sophisticated approach would be to include the directly measured voltage and a model-based state estimate to dynamically limit power so voltage stays within prescribed ranges. Along these same lines, a model-based dynamic limiter can be designed for any internal state that is desired to be within known ranges for safe and nondegrading operation.

Consider, for example, Li-ion chemistry, where our previously developed models can predict the physical limits of charge and discharge; namely, saturation/depletion of lithium concentration at the electrode surfaces $c_{s,e}$ and depletion of lithium concentration in the electrolyte solution c_e. To avoid sudden loss of power, the power must be limited to maintain lithium concentrations within constraints:

$$0 < \frac{c_{s,e}(x, t)}{c_{s,max}} < 1 \quad \text{and} \quad c_e(x, t) > 0. \tag{8.13}$$

To avoid damaging side reactions, the power must be limited to maintain the solid–electrolyte phase potential difference, $\phi_{s-e} = \phi_s - \phi_e$, within the constraints

$$U_{sd} < \phi_{s-e}(x, t) < U_{si}, \tag{8.14}$$

where U_{si} and U_{sd} are the equilibrium potentials of a side reaction that occurs when lithium ions are either inserted into or deinserted from active particles, respectively.

The dynamic power-limiting equations come directly from (8.11) and (8.12) with $\Delta t = 0$ because the desired limit is instantaneous. Substituting $\Delta t = 0$ into (8.9),

$$\mathbf{F} = \mathbf{I} \quad \text{and} \quad \mathbf{g} = \mathbf{0}, \tag{8.15}$$

so Equations (8.11) and (8.12) become

$$P_{min}(t) = \frac{V_{max} - \mathbf{c}^{\mathsf{T}}\mathbf{x}(t)}{d}, \tag{8.16}$$

$$P_{max}(t) = \frac{V_{min} - \mathbf{c}^{\mathsf{T}}\mathbf{x}(t)}{d}. \tag{8.17}$$

The dynamic power limits in Equations (8.16) and (8.17) depend on the measured state $\mathbf{x}(t)$. In practice, the full state is not measured, so the state estimate $\hat{\mathbf{x}}(t)$ must be used. If $d = 0$, meaning that there is no direct feed through from power input to the output of interest, then the power limits become infinite and meaningless. For voltage-based power limits, d is the cell resistance, which is always nonzero.

Figure 8.6 shows the block diagram for the dynamic power-limiting controller. The system produces the output voltage and $\mathbf{c}^{\mathsf{T}}\mathbf{x}(t)$. The differences with V_{min} and V_{max} are calculated and multiplied by $G = 1/d$ to form the upper and lower power limits, respectively. A single-sided pulse input is then saturated with the dynamic limits P_{max} and P_{min}.

Figure 8.7 shows the response to repeated charge pulses where the current is limited to ensure that $V(t) \leq V_{max}$. During the first charge pulse, the voltage stays under V_{max} so the current equals the desired current. The second charge pulse is clipped halfway through because the voltage limit has been reached. The voltage flatlines at the limit until the pulse ends. During ensuing pulses the voltage flatlines at the limit much more quickly, and after the third pulse

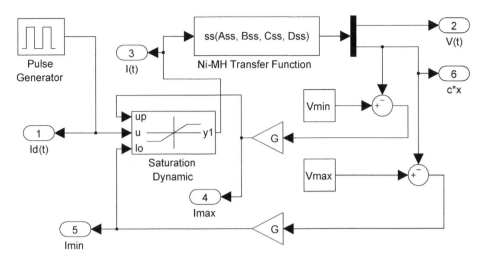

Figure 8.6 Dynamic power-limiting controller block diagram

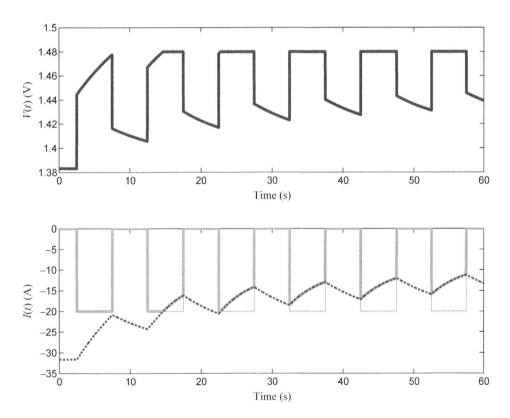

Figure 8.7 Dynamic power-limiting controller response: voltage and current (desired (thin), limited (thick), and dynamic limit (dashed))

the current never reaches its desired amplitude. In this linear simulation current and power are simply proportional to each other, so current limits are the same as power limits. For large power pulses with wide swings in voltage, however, the actual power will differ from the linear approximation.

Compared with empirical-based estimation algorithms and fixed power limits, dynamic, model-based power limits are complex and require accurate parameter and state estimation. A model-based approach is motivated by the fact that HEV battery packs, costing thousands of dollars per vehicle, are not used to their full power and energy capability because of the uncertainty about when damage or a sudden loss of power may occur in the real-time environment. This example demonstrate a practical method of controlling a battery to internal states and/or physicochemical constraints, thereby enabling an expanded range of power capability. In practice, the BMS could have multiple constraints on current, voltage, SOC, and so on and take the minimum at every time instant.

8.5 Pack Management

8.5.1 Pack Dynamics

Typical battery packs have many cells in series and parallel to provide the desired power. Figure 8.8 shows that cells in series share the same current and cells in parallel share the same voltage. If the cells are identical, then the voltage for a series string is $V = N V_j$, where N and V_j are the number of cells and their voltage, respectively. Identical cells in parallel all experience the same current $I_j = I/N$.

The voltage for a single cell is on the order of volts, so a device needs roughly as many cells in series as the desired maximum voltage. Individual cells are capacity and current limited. Placing cells in parallel multiplies the pack capacity and available current. Vehicle applications are high power, requiring many cells in series and parallel. Hundreds to thousands of cells are needed in an EV battery pack. Even a Pb–acid starter battery has six cells in series to provide 12 V output voltage and many cells in parallel to provide the cranking amps.

In this section, the previously developed cell models are combined to form pack models consisting of cells in series and parallel. The pack models can then be used to simulate the

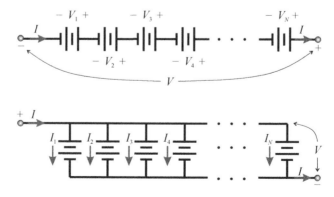

Figure 8.8   Battery cells in series (top) and parallel (bottom)

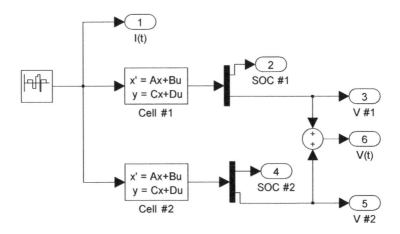

Figure 8.9 Block diagram for two cells in series

pack dynamics and design model-based controllers. For efficient numerical calculation, it is essential that the cell dynamics be distilled to the lowest order that accurately predicts the response. The focus here is on low-order, linear models that can be implemented on board as part of the BMS. Applications may require more sophisticated (e.g., nonlinear) models, however, depending on the desired precision of the simulation or model-based algorithm.

Series Strings

Figure 8.9 shows a block diagram for simulating two cells in series. The two-cell models are in state-space form with input $I(t)$ and outputs $SOC(t)$ and $V(t)$. The total voltage from the two cells in series equals the sum of the two cell voltages (output 6). The cell models can differ in initial conditions (e.g., different SOC(0)), capacity (different **B** matrices), and impedance (scale **D**).

Using an initial SOC = 70% for cell 1 and 65% for cell 2 produces the time response shown in Figure 8.10 with the pulse input current shown. The two cells are assumed to have identical dynamic models, so the 5% SOC error remains constant through the 25 min simulation. After an initial transient, the voltage of cell 2 tracks below that of cell 1, corresponding to the lower SOC. BMSs often use limits on SOC and/or voltage to determine when to stop charging or discharging. The individual cell voltages are known and measured. SOC can be estimated from voltage and current data. For these two cells in series, cell 2 reaches the lower SOC and voltage limits before cell 2 during discharge. Consider the SOC at $t = 10$ min. If the SOC limit was 55% then the current would have to be cut off at that time because cell 2 has reached the limit. Cell 1, on the other hand, still has 5% more capacity to provide that will not be used. Similarly, if the voltage limit was 1.34 V, then cell 2 crosses that line around $t = 10$ min and would trigger a current limiter. Again, cell 1 has charge and voltage to spare at that time that would not be utilized. The reverse is true on charge, with cell 1 triggering the SOC and voltage limits before cell 2. During discharge, cell 1 would not be fully utilized. During charge, cell 2 would not be fully charged. Thus, while the voltage provided by the two cells in series has almost doubled, their combined capacity has effectively been reduced due to cell imbalance.

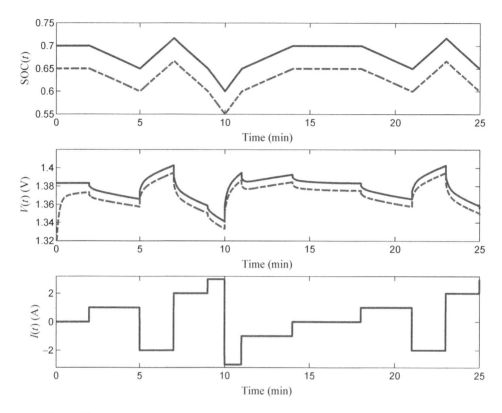

Figure 8.10 Simulation of two cells in series with different initial SOCs: SOC of cell 1 (solid) and cell 2 (dashed), voltage of cell 1 (solid) and cell 2 (dashed), and input current

Figure 8.11 shows the simulation results for two cells in series with different capacities. Cell 2 has a capacity that is 80% of cell 1. This may happen in practice due to differential aging and/or manufacturing variability. The cells start with the same initial SOC and voltage. Cell 2 rises and falls faster than cell 1 because it has 20% less capacity and the same current causes larger voltage swings. If the SOC stays fairly constant then this effect is minimized. If the SOC is steadily moving in one direction then the voltages and SOCs of the two cells deviate. In this case, cell 2 dictates when SOC or voltage limits are crossed during both charge and discharge. The smaller capacity cell 2 charges more quickly than cell 1 does; so, when the voltage/SOC cutoff has been reached by cell 2, cell 1 will not be fully charged. Similarly, cell 2 will not be fully discharged. The two cells in series provide twice the voltage but only the lower capacity of cell 1. The extra capacity available in cell 2 is not utilized.

In Figure 8.12 the two cells are identical with the same initial SOC except that cell 2 has 50% more resistance than cell 1. Impedance growth is typical of all battery chemistries as they age. Aging depends on the environment and initial manufacturing of each cell, so cell-to-cell variations inevitably magnify with cycle time. The SOC and capacity of the two cells are identical and the same current input is applied. The SOC for the two cells remains the same during the 25 min simulation because each cell has the same capacity and receives the same current. The voltage jumps due to resistance, however, are larger for cell 2. Thus, cell 2 will

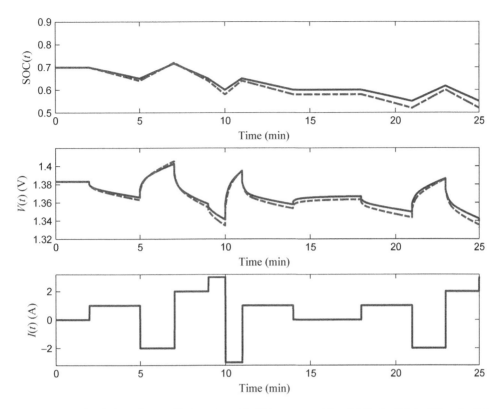

Figure 8.11   Simulation of two cells in series with different capacities: SOC of cell 1 (solid) and cell 2 (dashed), voltage of cell 1 (solid) and cell 2 (dashed), and input current

trigger voltage limits before cell 1, reducing the potential power of the two cells in series. Resistance differences between the two cells will not cause differences in SOC, however.

Differences between the cells in a series string can reduce the power and capacity of a series string relative to the power and capacity of the individual cells. SOC differences can be minimized by cell-balancing hardware in the BMS. Capacity loss manifests itself in SOC differences that can also be attacked by the cell-balancing hardware. Resistance changes, however, do not cause SOC differences, so they must be handled differently by the BMS. Although only two cells in series have been simulated here, these conclusions extend to series strings of arbitrary length.

Parallel Strings

For N cells connected in parallel, the overall impedance between input current $I(s)$ and output voltage $V(s)$ is

$$Z_p(s) = \left(\frac{1}{Z_1(s)} + \cdots + \frac{1}{Z_N(s)} \right)^{-1}, \tag{8.18}$$

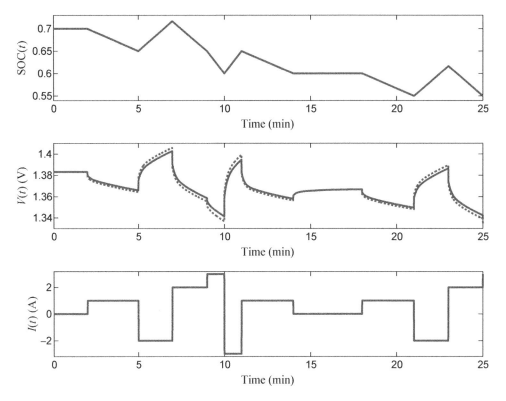

Figure 8.12 Simulation of two cells in series with different resistances: SOC of cell 1 (solid) and cell 2 (dashed), voltage of cell 1 (solid) and cell 2 (dashed), and input current

where $Z_1(s), \ldots, Z_N(s)$ are the individual cell impedances. Even if the individual cell impedances are of moderate order and N is large, then calculation of the overall impedance can be difficult.

For simulation purposes, the block diagram in Figure 8.13 can be used. The overall input $I(s)$ equals the sum of all the currents running through the parallel cells. The current through the first cell I_1 equals $I(s)$ minus the current $I_2, \ldots, I_N$. The top branch of the block diagram

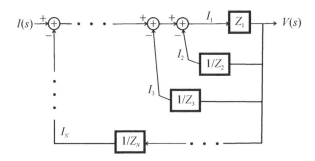

Figure 8.13 Block diagram for N cells in parallel

forms I_1 and, via the transfer function Z_1, $V(s)$. The feedback path with the block $1/Z_2$ produces the current I_2 from the voltage V. Each feedback loop in turn produces and other cell current and these are all subtracted from I to produce I_1. This block diagram conveniently provides the cell currents as signals to be exported.

Fortunately, cells have a direct feed-through resistance term that makes the numerator the same order as the denominator. Thus, both Z_j and $1/Z_j$ are proper transfer functions. If either transfer function had more zeros than poles then it would be improper because the output is obtained via differentiation rather than integration. Differentiation in real time can never be achieved because it requires information about future signals rather than signals from the past and current time. Also, although it is often true that the zeros of a cell's impedance transfer function are stable due to passivity [39], it is not necessary for this block diagram to produce a stable result. The block $1/Z_j$ may be unstable, but the closed loop is guaranteed to be stable because the cell impedance transfer functions are stable.

The overall impedance transfer function can be calculated using block diagram simplification as well. For a negative feedback block diagram with G in the forward path and H in the feedback path, the input/output transfer function

$$G_f = \frac{G}{1 + GH}. \tag{8.19}$$

Using this equation on the innermost feedback loop ($G = Z_1$ and $H = 1/Z_2$),

$$G_1 = \frac{Z_1}{1 + Z_1/Z_2} = \frac{Z_1 Z_2}{Z_2 + Z_1}, \tag{8.20}$$

or the impedance of two cells in parallel according to Equation (8.18). For the second feedback loop ($G = G_1$ and $H = 1/Z_3$),

$$G_2 = \frac{G_1}{1 + G_1/Z_3} = \frac{Z_1 Z_2 Z_3}{Z_2 Z_3 + Z_1 Z_3 + Z_1 Z_2}, \tag{8.21}$$

which equals the impedance of three cells in parallel. This pattern continues to the Nth cell so that the block diagram produces the proper impedance of N cells in parallel. The Matlab commands `feedback.m` and `series.m` can be used to form the parallel cell transfer functions given the individual cell impedances using this approach.

Figure 8.14 shows a Simulink model for two cells in parallel. The configuration in Figure 8.13 is used with $Z_1(s)$ in the forward path and $1/Z_2(s)$ in the feedback path. The total current and individual cell currents are output to the Matlab workspace. The second cell's transfer function is used to produce the SOC and voltage from the input I_2. For the case with identical cells, the current is equally divided between the two cells ($I_1(t) = I_2(t) = I(t)/2$). The voltage and SOC of the two cells track exactly.

Figure 8.15 shows the results when the capacity of cell 2 is 20% less than the capacity of cell 1. The cells start at 70% SOC and a series of 2 A discharge pulses is applied. The voltages of the two cells are identical, as required by the parallel connection. Cell 2 takes less current than cell 1, however, owing to its reduced capacity. Despite the lower current, the SOC of cell 2 decreases faster than cell 1 does, resulting in an SOC mismatch. After five pulses

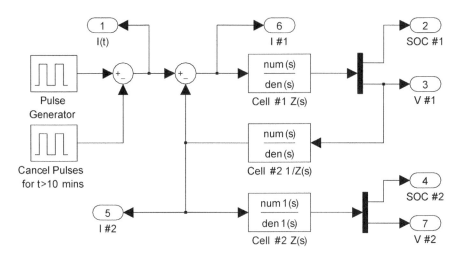

Figure 8.14 Block diagram for two cells in parallel

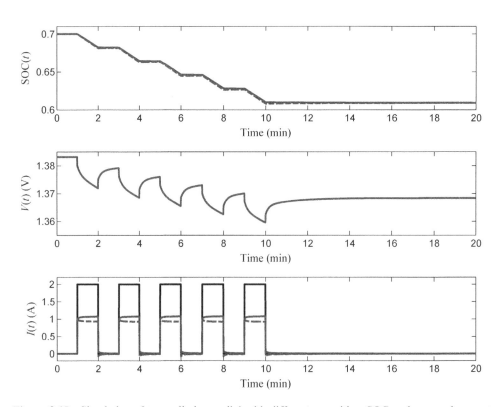

Figure 8.15 Simulation of two cells in parallel with different capacities: SOC, voltage, and current (cell 1, solid; cell 2, dashed; total current, solid)

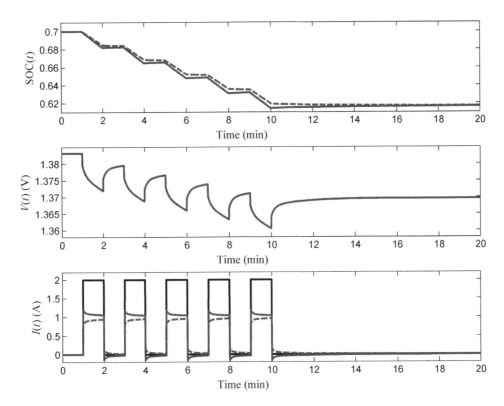

Figure 8.16 Simulation of two cells in parallel with different impedances: SOC, voltage, and current (cell 1, solid; cell 2, dashed; total current, solid)

(10 min), the pulses stop and the two cells equilibrate. The last 10 min of the simulation demonstrate the automatic equalization of cells in parallel, even if they are not identical. There is no SOC drift between cells in a parallel string, so there is no need for balancing. All cells have the same voltage, so voltage-based current limits would not be affected by cell variability. Other current limits based on internal cell variables, however, could be influenced by cell variations. The SOC and current are not necessarily the same for cells in parallel. Current is not evenly distributed because the stronger cell takes more current. The weaker cells show greater SOC variations for this example.

Figure 8.16 shows two cells of differing impedance under the same pulse discharge current input. Cell 2 has 50% more resistance than cell 1 does. Thus, for the same voltage, cell 2 has less current than cell 1 does. Unlike the previous case, however, the capacities are the same, so the SOC of cell 2 decreases at a slower rate than cell 1 does. Again, after the pulses cease, the cells equilibrate, showing automatic balancing. The final SOC and voltage are higher than the previous case because the capacity of the two cells in parallel is twice the single cell capacity, whereas cell 2 does not have full capacity in the previous case. Again, cell variability does influence the performance and potential limits on SOC and current, at least during and shortly after current pulses.

8.5.2   Cell Balancing in Series Strings

Cells in series strings naturally become unbalanced and remain so unless corrective action is taken. Some chemistries can rely on trickle charging and side reactions that occur in overcharged cells. These side reactions absorb current without increasing SOC to passively balance the pack. During trickle charging the fully charged cells will remain at 100% SOC while the undercharged cells slowly come up to 100% SOC. This method of cell balancing requires no additional hardware/software to implement. It can only be used, however, for chemistries that exhibit high SOC side reactions (e.g., Pb–acid and Ni–MH), not Li-ion cells. These side reactions can be detrimental to the health of the battery and increase aging. Finally, this approach only works for systems with periodic full charging (e.g., EVs but not HEVs). For safety and efficiency, most battery packs use some form of controlled balancing/equalization with switches and passive electrical components (resistors, capacitors, and inductors) or active DC–DC converters as discussed in this section.

Power Electronics

The power electronics for cell balancing range from relatively simple shunt resistors to complex circuits with buck-boost and/or flyback converters [85]. Equalization circuits can be classified as passive or active and series, parallel, or hybrid. All equalizers use MOSFETs, which are high-power electrical switches that control current flow. Passive cell-balancing methods discharge overcharged cells through shunt resistors. This low-cost approach has been implemented in, for example, the TI bq76PL536 chip shown in Figure 8.2. This method is highly inefficient, however, because the excess energy from the overcharged cells is dissipated as heat. Active cell balancing uses high-frequency switching and capacitors and/or inductors to transfer energy. This approach has the advantage of higher efficiency with the associated higher complexity and parts count. Series balancers transfer current cell to cell in the pack while a parallel configuration draws current from the pack bus voltage supply.

Figure 8.17 (left) shows a schematic diagram of a passive, series balancing shunt resistor circuit. The N battery cells are connected in series. A shunt resistor R_i and MOSFET switch S_i is connected in parallel with each cell. If the switch is turned off then the pack current I passes through the cell. If the switch is turned on then pack current is diverted through the resistor. The cell and the resistor are in parallel, so the total current $I = I_{si} + I_{ci}$ is the sum of the shunt resistor current I_{si} and the cell current I_{ci}. If the shunt resistance is small, then most of the current will flow through the shunt. The shunt discharges the cell through the resistor, dissipating energy and decreasing the efficiency of the pack. It also is only useful for balancing during charge.

The simplest active balancing circuit is the switched capacitor design shown schematically in Figure 8.17 (right). This is also a series balancer, because current is moved between cells in the pack, not between the cells and the bus. Each cell has a single pole, double-throw switch and a capacitor. Other switched capacitor balancers include resistors or lump the capacitance in a single capacitor for the entire pack [86]. The principle of operation is shown in Figure 8.18. If cell k has a higher voltage and, hence, higher SOC than cell $k + 1$ and the switches are in their lower position, then capacitor C_k charges up to voltage V_k. When the switches move to their upper position, the capacitor discharges through cell $k + 1$, raising its voltage and SOC. The rate of current transfer decreases as the two cell voltages converge.

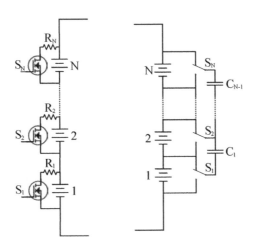

Figure 8.17 Schematic diagrams of a shunt resistor balancing circuit (left) and a switch capacitor balancing circuit (right) [85]

The switched capacitor balancing system can be operated continuously without supervisory control, so it is relatively low cost and easy to implement. Cells with low voltage can be brought up and cells with high voltage can be reduced. The equalization power is limited, however, so it may take a long time to balance the system, especially for chemistries with flat voltage versus SOC curves. With severely mismatched cells, a switched capacitor balancing system may not be able to provide enough current to equalize cell voltages.

Figure 8.19 shows a schematic diagram of a bidirectional flyback converter balancing circuit. The part count and complexity of this design are significantly higher than the previous two examples, requiring MOSFETs, diodes, and transformers for each cell. This complexity is required because current is drawn from the high-voltage bus to balance the individual cells in the string. The bidirectional design allows the transfer of current to and from the cells to the bus. Higher currents can be used, so equalization time can be reduced. The equalization circuit design provides higher control authority for tight regulation of cell voltage and SOC.

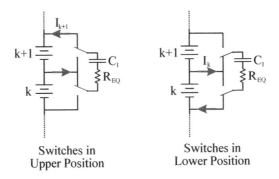

Switches in
Upper Position

Switches in
Lower Position

Figure 8.18 Schematic diagram of the switched capacitor balancing circuit [85]

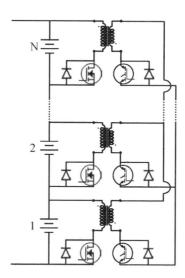

Figure 8.19   Schematic diagram of the bidirectional flyback converter balancing circuit [85]

Pulse Width Modulation

Cell-balancing circuits involve switches that are typically implemented using power MOSFET transistors. Control algorithms usually are not on–off or bang–bang controllers but require continuously variable control inputs. One can approximate a continuously variable signal with a switched signal using pulse width modulation (PWM). The desired, continuously varying signal is approximated by an on–off pulse train that has a fixed period corresponding to the PWM frequency. The on time is adjusted relative to the overall period, so the average value of the PWM output is the same as the continuous signal. The on time divided by period is called the duty cycle of the pulse width modulator. The plant, in this case the battery cell, is assumed to have a low-pass filtering effect that reduces the high-frequency content at the PWM frequency and above. In this case, only the low-frequency, average signal remains, and the cell responds to the PWM signal as it would have responded to the continuous signal. A low-pass filter is often added to further smooth the PWM output.

A simple PWM algorithm is shown in Figure 8.20 as the shaded blocks in the block diagram. The PWM frequency is chosen to be 100 Hz, much slower than is typically implemented in practice. In numerical simulation, however, higher frequencies require shorter integration times and longer run times. The 100 Hz PWM frequency chosen for this example is much faster than the dynamics of the Ni–MH cell model, which has a cutoff frequency around 10 Hz. Cycling frequencies less than 10 Hz have been found to reduce the life of Li-ion cells [87].

The triangle wave ramps between zero and one and resets to zero after 0.01 s. The PWM output is either one or zero associated with the switch closed or open. The PWM input is compared with the triangle wave. The "Compare to Zero" block outputs a one when the triangle wave is less than the PWM input and zero when the triangle wave is greater than the PWM input. If the PWM input is less than zero, the triangle wave is greater than the PWM input for the entire period, so the PWM output is zero (duty cycle equal to 0%). If the PWM

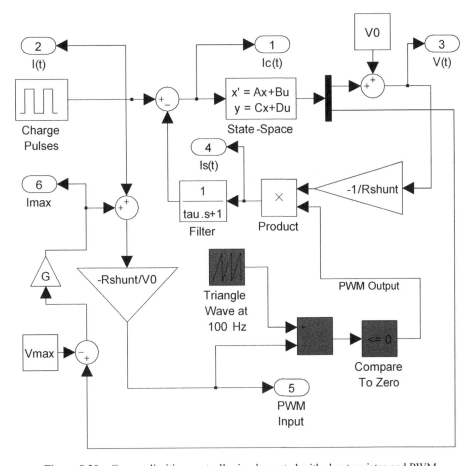

Figure 8.20   Current-limiting controller implemented with shunt resistor and PWM

input is larger than one then the triangle wave is always less than the PWM input and the output stays at one (duty cycle equal to 100%). PWM inputs between zero and one have duty cycles that equate the average values of the PWM input and output.

Shunt Resistor Equalization System

The instantaneous pulse power-limiting controller developed in Section 8.4 can be implemented using a shunt resistor balancing circuit. The shunt transistor is switched on and off to limit the current such that the voltage stays at or below a specific limit V_{max} during charge. This will allow the cell to reach the maximum voltage and 100% SOC but not go into overcharge. The cell model in state variable form is

$$\dot{x}(t) = \mathbf{A}x(t) + \mathbf{b}I_c(t) \tag{8.22}$$

$$V(t) = \mathbf{c}^T x(t) + dI_c(t) + V_0, \tag{8.23}$$

where $I_c(t)$ is the cell input current, V_0 is the equilibrium voltage at the linearization operating point, and $V(t)$ is the output voltage. The block diagram in Figure 8.20 shows that the cell input current $I_c = I - I_s$, where $I(t)$ is the pack input current and

$$I_s(t) = \frac{-V(t)}{R_{shunt}} \tag{8.24}$$

is the shunt resistor current. The maximum cell current is calculated from

$$I_{c,max} = \frac{V_{max} - V_0 - \mathbf{c}^T\mathbf{x}(t)}{d}. \tag{8.25}$$

The required shunt current to limit I_c to $I_{c,max}$ is calculated from

$$I_s = I - I_{c,max}, \tag{8.26}$$

normalized by $-R_{shunt}/V_0$ to provide an input signal nominally between zero and one, and sent to the pulse width modulator. The product block acts as the MOSFET switch. If the PWM output is one then the output of the product block is $-V/R_{shunt}$. If the PWM output is zero then the output of the product block is zero. PWM noise in the $I_s(t)$ signal is reduced using a unity gain, low-pass filter with a cutoff frequency of 5 Hz. The filtered shunt current subtracts from the pack current to limit the current flowing into the cell.

Figure 8.21 shows the simulation results for the current-limiting controller implemented with a shunt resistor and PWM. A square-wave pulse charge input between 0 and -20 A is input to the cell. The cell voltage limit is set at 1.48 V. During the first pulse, the voltage stays below the cutoff, so the current pulse is above the cutoff and the shunt resistor remains off. The PWM input is below zero during this time, so the PWM output is also zero. During the second pulse, however, the cell voltage reaches V_{max}. The PWM input becomes positive and the PWM output switches on and off at an increasing duty cycle until the pulse ends. This holds the cell voltage at approximately the limit because I_s is taking some of the charging current, reducing I_c below $I_{c,max}$. For subsequent pulses, more and more of the current flows through the resistor shunt instead of the cell. During the last pulse, the switch remains fully on at a 100% duty cycle. This is not enough shunt current, however, to prevent the cell voltage from increasing above V_{max}. The overall string current must be limited at this point to prevent the cell from overcharging.

The PWM snapshots indicate that the PWM algorithm is converting the PWM input to an on–off PWM output with a proportional duty cycle. From $t = 14.7$ s to 14.8 s, the 10 cycles of the PWM are at a low duty cycle with the switch only closed for a short part of the cycle. During the intermediate time period, the PWM output shows a larger duty cycle of roughly 80%. At the end of the simulation, the duty cycle becomes 100% as the PWM input crosses one.

Including the PWM in the simulation model provides accurate results, including the switching noise propagation through the system and filter dynamics, but complicates the numerical solution. The simulation is nonlinear and the high-frequency switching requires a small integration time step, increasing the run time and introducing instabilities in the numerical code.

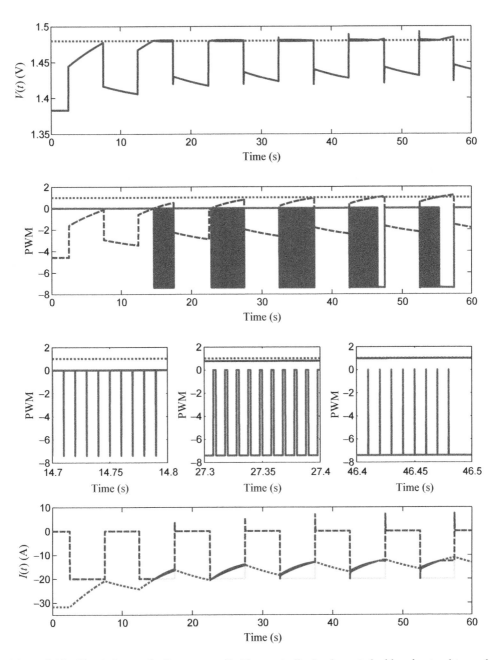

Figure 8.21 Simulation results for a current-limiting controller implemented with a shunt resistor and PWM: cell voltage (solid) and maximum limit (dotted); PWM input (dashed), output (solid), and full on condition (dotted); PWM snapshots at different 0.1 s time windows; and input current (light solid), cell current (dashed), and cell current limit (dotted)

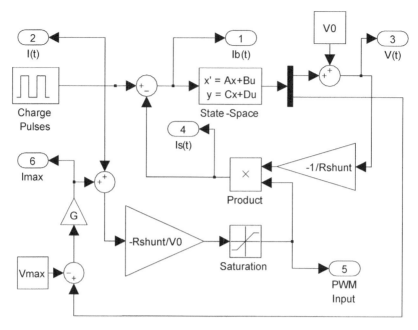

Figure 8.22   Current-limiting controller implemented with shunt resistor (PWM dynamics neglected)

The objective of the PWM is to generate on–off approximations of continuous signals. If we neglect the PWM dynamics, then the simulation can run much more quickly and the results may be sufficiently accurate. Figure 8.22 shows the previous system with the PWM dynamics neglected. The output of the $-R_{shunt}/V_0$ block is saturated between zero and one. The signal entering the product block is no longer zero or one, but between zero and one. The filter is no longer required to remove the PWM noise. Figure 8.23 is remarkably similar to Figure 8.21. The signals are cleaner without PWM noise contamination. The spike transients associated with the noise filter are gone. The shunt current signals are smooth curves rather than discontinuous PWM. The cell voltage and current responses show the same characteristic responses. Thus, continuous approximations of the PWM or neglect of PWM dynamics is often reasonable for cell-balancing system analysis.

Switched Capacitor Equalization System

Figure 8.24 shows a schematic diagram of four MOSFETs switches being used to equalize two cells in a series string [88]. This hardware configuration can be repeated as needed to equalize all of the cells in a series string. The two cells are shown with voltage V_1 and V_2. The three capacitors used in the circuit help filter the switch transients (C_{filter}) and store energy (C_{EQ}). The equalization capacitor C_{EQ} enables the transfer of current from the cell with higher voltage to the cell with lower voltage. The schematic on the left of Figure 8.24 shows the current flow if $V_2 > V_1$ with Drive 1 set high (turning on the first Q_1 and third Q_3 MOSFET switches) and Drive 2 set low (turning off the second Q_2 and fourth Q_4 MOSFET

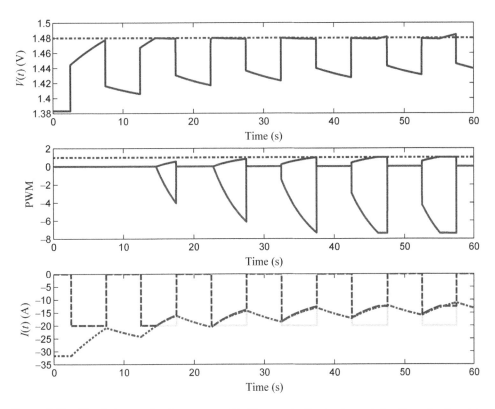

Figure 8.23 Simulation results for a current-limiting controller implemented with a shunt resistor (PWM dynamics neglected): cell voltage (solid) and maximum limit (dotted); PWM input (dashed), shunt current (solid), and full on condition (dotted); and input current (light solid), cell current (dashed), and cell current limit (dotted)

switches). Current flows from the positive terminal of the second cell, through Q_1 and into the equalization capacitor. Current then returns to the negative terminal of the second cell through Q_3. Once the equalization capacitor is charged, Drive 1 is set to high and Drive 2 set to low, turning off Q_1 and Q_3 and turning on Q_2 and Q_4. Current flows from the equalization capacitor through Q_2 and charges the first cell. The current then flows back to the capacitor through Q_4.

The switching of Drive 1 and Drive 2 is done at a fixed, high frequency. The equalization process happens automatically without the need for sensing or control. Current flows passively from higher voltage cells to lower voltage cells. If the cells are at the same voltage then no current flows between them. For long series strings, the flow is similarly passive and automatic from high-voltage cells to low-voltage cells.

The high-frequency switching averages out to act as a parallel resistor that connects the two cells and allows them to balance without reducing the series voltage. To determine the effective resistance of this parallel connection, the time response of the switched capacitance circuit is calculated. When a cell is connected to the capacitor, the circuit consists of a voltage

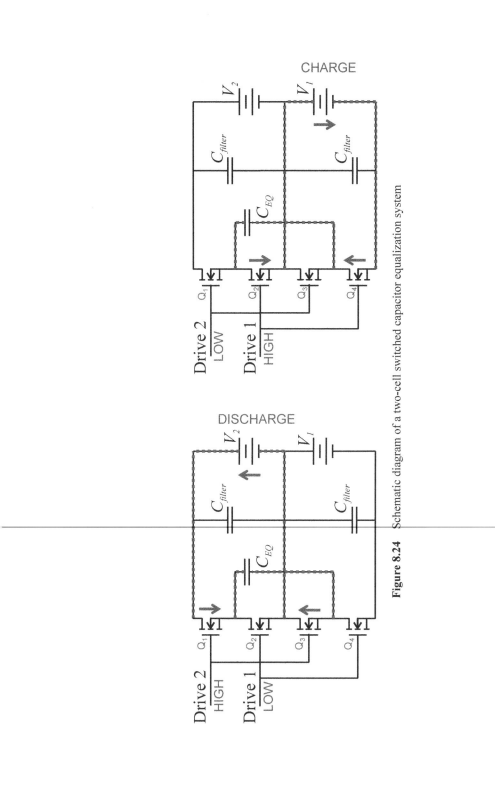

Figure 8.24 Schematic diagram of a two-cell switched capacitor equalization system

source $V_b(t)$, a resistor R_{EQ} that models the physical connections, the switch resistance, and the capacitor equivalent series resistance, and the equalization capacitor C_{EQ}. The governing differential equation in terms of the capacitor voltage $V_c(t)$ is

$$\dot{V}_c = \frac{1}{R_{EQ}C_{EQ}}(V_b - V_c), \tag{8.27}$$

with initial condition V_0. If the battery voltage is assumed constant, then the solution of Equation (8.27) is

$$V_c(t) = V_b + (V_0 - V_b)\,e^{-t/R_{EQ}C_{EQ}}. \tag{8.28}$$

The constant cell-voltage assumption is valid because the switching frequency is high (kilohertz) relative to the cell bandwidth (hertz). The current transferred between the cell and the capacitor is

$$I_c(t) = \frac{V_b - V_c}{R} = \frac{V_b - V_0}{R}\,e^{-t/R_{EQ}C_{EQ}}. \tag{8.29}$$

If Drives 1 and 2 are set high and low, respectively, then $V_b = V_2$ and $V_0 = V_k$, where V_k is the capacitor voltage at the beginning of the kth switch cycle. The capacitor voltage is continuous during switches but the current can jump. After the capacitor has been connected to V_2 for Δt seconds, Drives 1 and 2 are set to low and high, respectively. The solutions in Equations (8.28) and (8.29) can be used in this case as well with $V_b = V_1$ and $V_0 = V_{k+1}$, where V_{k+1} is the capacitor voltage at the middle of the kth cycle. Combining the solutions for the two switching conditions, we have

$$V_{k+1} = V_2 + (V_k - V_2)\,e^{\Delta t/R_{EQ}C_{EQ}}, \tag{8.30}$$

$$V_k = V_1 + (V_{k+1} - V_1)\,e^{\Delta t/R_{EQ}C_{EQ}}, \tag{8.31}$$

where the capacitor voltage is assumed periodic ($V_c(t + 2\Delta t) = V_c(t)$) because the cell voltages are assumed constant and the amount of charge provided to the capacitor by the high-voltage cell must equal the amount of charge supplied by the capacitor to the low-voltage cell in steady state. The average current provided to the capacitor is

$$I_{ave} = \frac{1}{\Delta t}\int_0^{\Delta t} I_c(\tau)\,d\tau = \frac{C}{\Delta t}[V_b - V_0 + (V_0 - V_b)\,e^{\Delta t/R_{EQ}C_{EQ}}]. \tag{8.32}$$

Solution of Equations (8.30) and (8.31) yields

$$V_{k+1} = \frac{e^{\Delta t/R_{EQ}C_{EQ}}V1 + V2}{e^{\Delta t/R_{EQ}C_{EQ}} + 1}, \tag{8.33}$$

$$V_k = \frac{e^{\Delta t/R_{EQ}C_{EQ}}V2 + V1}{e^{\Delta t/R_{EQ}C_{EQ}} + 1}. \tag{8.34}$$

Substitution of these equations into Equation (8.32), yields the average current

$$I_{ave} = \frac{C(V_2 - V_1)(1 - e^{\Delta t / R_{EQ} C_{EQ}})}{\Delta t (e^{\Delta t / R_{EQ} C_{EQ}} + 1)}.$$

(8.35)

This equation shows that the current shuttled between the two cells is proportional to their voltage difference. The equivalent resistance

$$R_{equiv} = \frac{\Delta t (e^{\Delta t / R_{EQ} C_{EQ}} + 1)}{C(1 - e^{\Delta t / R_{EQ} C_{EQ}})},$$

(8.36)

or in nondimensional form

$$\frac{R_{equiv}}{R_{EQ}} = \frac{\tau(1 + e^{-\tau})}{1 - e^{-\tau}}.$$

(8.37)

The equivalent resistance has a minimum value of $2R_{EQ}$ at $\tau = \Delta t / R_{EQ} C_{EQ} = 0$ (infinitely high frequency switching). As τ gets large, $R_{equiv} \approx \tau R_{EQ}$.

For fast balancing, R_{equiv} should be as small as possible so that a small voltage difference will produce a large balancing current. The minimum R_{equiv}, however, is limited by the parasitic resistance in the connections, switches, and capacitors, which cannot be entirely eliminated. The switching frequency can be chosen very high (Δt small) and/or C_{EQ} chosen large to approach this minimum R_{equiv}. If $\tau < 1$ then $R_{equiv} \leq 2.16 R_{EQ}$ and within 16% of its minimum value.

Figure 8.25 shows the time response of the two-cell equalizer shown in Figure 8.24, where the cell voltages are $V_2 = 1.48$ V and $V_1 = 1.4$ V. The parasitic resistance is set at a typical value of $R_{EQ} = 0.5$ Ω. Two cases are run with different capacitance and switching frequency. The first case with $C_{EQ} = 100$ μF and $f = 10$ kHz, shown by the gray curves, corresponds to $\tau = 1$. The second case ($C_{EQ} = 50$ μF and $f = 5$ kHz – black curves) corresponds to $\tau = 4$. For both cases the capacitor charges during the first half of each cycle and then discharges during the second half. The current alternates between positive current flowing from discharge of cell 2 into the capacitor and negative current flowing to charge cell 1. The steady-state voltages calculated by Equations (8.33) and (8.34) (dashed lines) match the actual voltages at the ends and middle of each cycle. The second case allows more complete charge of the capacitor, so the steady-state voltages approach V_1 and V_2 at the ends and middle of each cycle, respectively. The first case involves a smaller voltage swing on the capacitor, but the average current is larger. The average voltage in both cases is the same, so a larger average current means that more power is being transferred from cell 2 to cell 1. The average voltage divided by average current matches the R_{equiv} calculated in Equation (8.37).

The high-frequency cycling of the switched-capacitor equalizer can be approximated by a parallel resistor of resistance R_{equiv} connected between the two cells, as shown in the Simulink block diagram in Figure 8.26. The current flowing into the cells is reduced by the voltage difference divided by R_{equiv}. The simulation results for two cases are shown in Figure 8.27. where the Ni–MH cells have either mismatched initial SOC or capacity. For the initial SOC mismatch case, the first cell is given an initial SOC of 65% and the second cell is initially at 70%. The solid curves show that over time the SOCs of cells 1 and 2 converge. In this case, a

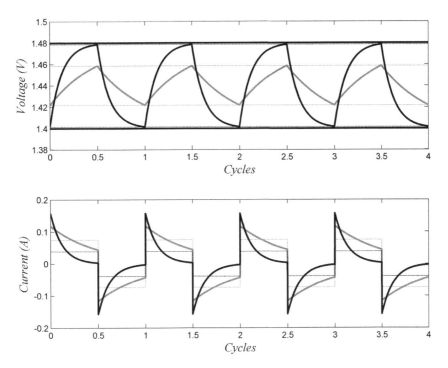

Figure 8.25 Switching response of two-cell equalizer in Figure 8.24 with $V_2 = 1.48$ V and $V_1 = 1.4$ V: capacitor voltage and current versus cycles for $R_{EQ} = 0.5$ Ω and $C_{EQ} = 100$ μF and $f = 10$ kHz (gray curves) and $C_{EQ} = 50$ μF and $f = 5$ kHz (black curves)

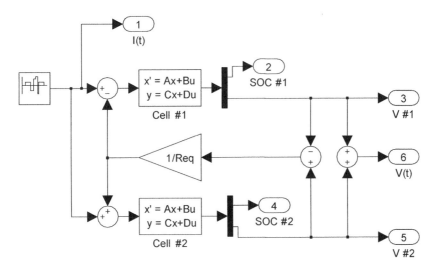

Figure 8.26 Two-cell switched capacitance equalization Simulink block diagram

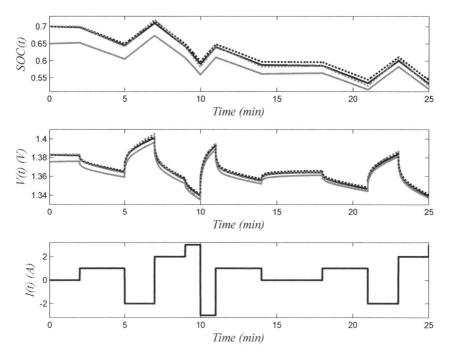

Figure 8.27 Switched-capacitor equalization simulation: SOC, cell voltages, and input current (cell 1, dark; cell 2, light) for 5% initial SOC mismatch (solid) and 20% capacity mismatch (dashed)

very small $R_{\mathrm{equiv}} = 0.1\,\Omega$ is used. This may not be physically realizable, but it demonstrates convergence in this short simulation time of 25 min. It is interesting to note that the voltage converges much more rapidly than SOC does. If the cells are otherwise identical then voltage and SOC convergence is guaranteed. In the second case (dashed lines), the initial SOCs are identical at 70% but the capacity of cell 1 is 20% less than cell 2. This causes the voltage and SOC to drift apart, especially under extended unidirectional current charge/discharge. The dashed curves steadily move apart despite the high-gain R_{cquiv}, indicating the potential for SOC divergence despite cell equalization. Given enough time without current charge/discharge to the string, however, the voltages and SOC will eventually converge. This simulation points out the limitations of switched capacitance for mismatched series strings with relatively high parasitic resistance. This problem is further exacerbated in long series strings due to the slow diffusion of current along the string [88].

8.5.3 Thermal Management

The goal of a battery thermal management system is twofold: managing cell temperatures in a preferred range of 5–40 °C under all environmental and operating conditions and maintaining cell-to-cell temperature difference within 3–5 °C. This entails cell cooling in warm environments to remove heat generated from battery operation and cell heating in very cold climates.

Cell cooling is important to ensure the life and safe operation of the cells. Cells tend to degrade more rapidly at high temperatures. While long life is important, the safety of a battery pack is paramount. At 175 Wh/kg, the electrical energy density of an Li-ion cell, for example, is roughly 15% of that of common explosives such as TNT ($\approx$1160 Wh/kg). Thus, if all electrical energy were released from an Li-ion cell as heat, there would be significant damage done. An electric vehicle with a driving range of 200–300 miles per charge would need 400 Wh/kg batteries. If these advanced batteries are indeed successfully developed and deployed in transportation, their electrical energy density would then amount to 30% of that of explosives, implying that battery safety would be a tremendous challenge for these vehicles.

Cell cooling is needed to remove the heat generated during battery operation due to reversible entropic heat ($T \Delta S$) associated with electrode reactions, Joule heating ($I^2 R$), and irreversible reaction heat ($I \eta$). A detailed elaboration of various heat sources and their physical origins is provided in Chapter 3 and [89]. To summarize, the total heat generation rate can be approximated as

$$q(t) = I \left(U - V - T \frac{\partial U}{\partial T} \right), \tag{8.38}$$

where U is the open-circuit potential and T is cell temperature. The temperature dependence of the electrochemical properties that dictate battery performance is usually described by the Arrhenius equation introduced in Chapter 3 and repeated here for convenience:

$$\Psi = \Psi_{\text{ref}} \exp \left[\frac{E_{\text{act}}^{\Psi}}{R} \left(\frac{1}{T_{\text{ref}}} - \frac{1}{T} \right) \right], \tag{8.39}$$

where Ψ_{ref} is the property value defined at the reference temperature $T_{\text{ref}} = 25°C$. The activation energy E_{act}^{Ψ} has units of joules per mole and controls the temperature sensitivity of each individual property, Ψ. Important physiochemical properties include the solid- and liquid-phase diffusion coefficients, electrolyte ionic conductivity, and exchange current densities of electrode reactions.

According to Ji et al. [90], the activation energy for these electrochemical properties typically ranges from 30 to 68 kJ/mol, translating to roughly doubling of the reaction rate and hence doubling of the heat generation per Equation (8.38) for every 10 °C temperature rise. On the other hand, the heat dissipation rate increases only linearly with the battery temperature rise, assuming a constant heat transfer coefficient from battery surfaces to the ambient. This is the fundamental reason why lithium batteries are prone to catastrophic thermal runaway.

In addition to avoidance of thermal runaway, uniformity of temperature is very critical for a battery module or pack consisting of multiple cells. To appreciate this importance, consider cells connected in parallel. If the cell-to-cell temperature difference is between 3 and 5 °C, the currents passing through individual cells would differ by 25–40% under the same voltage. Obviously, these cells will develop significant imbalance over multiple charge and discharge cycles, leading to premature failure or shortened cycle life.

There are several approaches to battery thermal management: using air, liquid, refrigerant, or by phase-change materials. Figure 8.28 illustrates examples of cooling/heating by cabin air and dedicated liquids. For liquid cooling, the working fluid passes through a cooling jacket around the cells, is heated using the vehicle engine coolant, and is cooled using the air conditioner. In air cooling, temperate cabin air is often used to improve heating/cooling

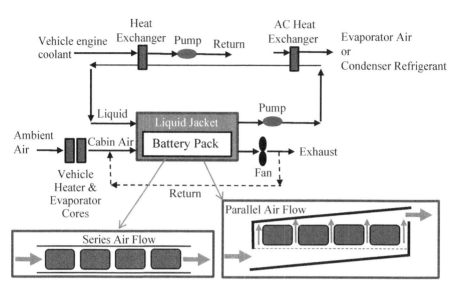

Figure 8.28 Schematic of liquid (top) and air (bottom) cooling and heating systems with series and parallel flow paths (insets)

efficiency. The design of the fluid cooling channels in the pack can include series, parallel, and fin-type architectures. The insets in Figure 8.28 depict schematics of series and parallel air flow for heating/cooling.

Startup of an electric vehicle, and hence a battery pack, from subzero temperatures requires fast heating of batteries. This is because at low temperatures, such as $-20°C$, most battery chemistries exhibit drastically reduced performance. In Li-ion cells, for example, charge at low temperatures may cause substantial degradation such as lithium plating in the graphite anode. Rapid heating of batteries from subzero temperatures is a topic of increasing practical significance [91] and can be achieved using:

1. **Self-heating.** Heat is generated within the battery via specially designed current-pulsing profiles.
2. **External heating.** An external electrical heater and a closed heating loop transfer electrical energy from the battery to a heater (discharge), and then thermal energy from the heater to the battery through a heat-transfer medium (e.g., air) in the loop.
3. **Enhance external heating.** A heat pump and a closed heating loop produce more heat than the electrical energy discharged from the battery, thereby achieving higher heating efficiency.

The detailed design of a battery thermal management system can be achieved using coupled electrochemical–thermal models based on the governing equations of Chapter 3 including temperature dynamics and temperature-dependent model parameters. Commercial codes can also be used for design and simulations studies, including the computer-aided tool called AutoLion^{TM} developed by EC Power (www.ecpowergroup.com). Using this software, a 2.8 kWh battery pack shown in Figure 8.29, consisting of 24 cells of 35 Ah with two in parallel and 12 in series, is simulated together with its air cooling system. Figure 8.30 displays

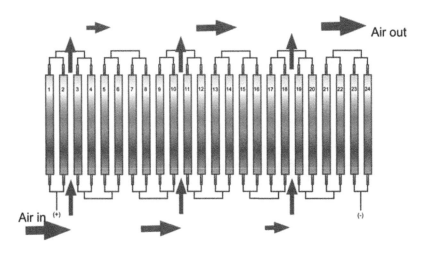

Figure 8.29 Schematic diagram of a 2.8 kWh battery pack with air cooling

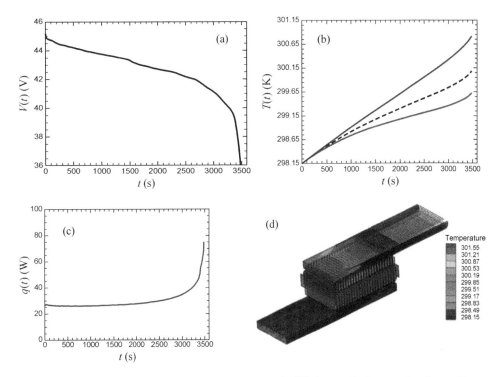

Figure 8.30 Simulation results using EC Power AutoLion™ for 1C discharge: (a) voltage, (b) temperature (max, top; average, middle; min, bottom), (c) heat generation, and (d) temperature distribution after discharge

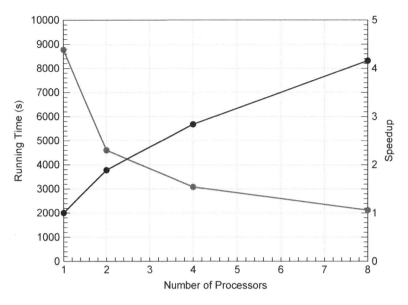

Figure 8.31 Computation time (left axis) and speedup (right axis) for EC Power AutoLion™ versus number of processors

the simultaneous predictions of the battery $1C$ discharge curve, temperature evolution, heat generation evolution, and three-dimensional temperature field at the end of $1C$ discharge [92]. The computational time is quite reasonable. With a single processor, this simulation takes 8800 s for the $1C$ discharge process. However, if using four CPUs on a single desktop, the simulation time drops to 3000 s, which is faster than the real testing time (3600 s). Figure 8.31 displays the computational time and speedup with more processors. It is evident that complete simulation of battery packs is feasible and highly efficient with relatively inexpensive computer hardware. It has been demonstrated that a coupled electrochemical–thermal simulation of a full-scale EV battery pack (20 kWh) takes only a fraction of the real testing time if using 64 CPUs.

Problems

8.1 A state-space model of a 8.75 h Pb–acid cell, linearized at 70% SOC (2.12 V), has the state matrices

$$\mathbf{A} = \begin{bmatrix} 0 & 0 & 0 \\ 0.5113 & -0.01005 & 0 \\ 0 & 0 & -0.2727 \end{bmatrix}, \quad \mathbf{b} = \begin{bmatrix} 1.332 \times 10^{-6} \\ 0.0002907 \\ -0.002907 \end{bmatrix}, \quad (A.8.1)$$

$$\mathbf{c}^{T} = \begin{bmatrix} 0 & 1 & -1 \\ 23.83 & 0 & 0 \end{bmatrix}, \quad \text{and} \quad \mathbf{d} = \begin{bmatrix} 0.005 \\ 0 \end{bmatrix}, \quad (A.8.2)$$

where the first output is the voltage $V(t)$, the second output is SOC, and the input is the current $I(t)$.

(a) Simulate the response to a square wave input with 1 A amplitude and a 100 s period for six complete cycles of two cells in series with initial SOCs of 0.65 and 0.7. Plot the cell responses of the voltage, SOC, and current on the same plots.

(b) Simulate the response as in part (a) for two cells in parallel. Plot the cell responses of the voltage, SOC, and current on the same plots.

(c) Simulate the response as in part (a) for two cells in parallel with the same initial SOC but cell 2 has twice the contact resistance. Plot the cell responses of the voltage, SOC, and current on the same plots.

(d) Simulate the response as in part (a) for two cells in parallel with the same initial SOC but cell 2 has 10% less capacity. Plot the cell responses of the voltage, SOC, and current on the same plots.

(e) Simulate the response as in part (a) for two cells in series with a switched-capacitance equalization circuit. Use $R_{EQ} = 0.5\ \Omega$, $C_{EQ} = 100\ \mu F$, and $f = 10$ kHz. Plot the cell responses of the voltage, SOC, and current on the same plots.

8.2 Using the 6 Ah Li-ion cell transfer function from Eq. (A.7.1), design a dynamic current limiter that keeps the cell voltage above 3.55 V.

(a) Assume the full state is known and simulate the response to the duty cycle input in Figure 6.11. Plot the voltage and current response.

(b) Design an observer to estimate the state based on current and voltage measurements and use the estimated state in the dynamic current limiter. Simulate the response to the duty cycle input in Figure 6.11. Plot the voltage and current response.

References

[1] Pistoia, G. (2009) *Battery Operated Devices and Systems: From Portable Electronics to Industrial Products*, Elsevier, New York, NY.

[2] Linden, D. and Reddy, T.B. (2002) *Handbook of Batteries*, McGraw-Hill, New York, NY.

[3] Schiffer, J., Sauer, D.U., Bindner, H. *et al.* (2007) Model prediction for ranking lead–acid batteries according to expected lifetime in renewable energy systems and autonomous power-supply systems. *Journal of Power Sources*, **168** (1), 66–78.

[4] Ruetschi, P. (2004) Aging mechanisms and service life of lead–acid batteries. *Journal of Power Sources*, **127** (1–2), 33–44.

[5] Taniguchi, A., Fujioka, N., Ikoma, M., and Ohta, A. (2001) Development of nickel/metal-hydride batteries for EVs and HEVs. *Journal of Power Sources*, **100** (1–2), 117–124.

[6] Bäuerlein, P., Antonius, C., Löffler, J., and Kümpers, J. (2008) Progress in high-power nickel–metal hydride batteries. *Journal of Power Sources*, **176** (2), 547–554.

[7] Wohlfahrt-Mehrens, M., Vogler, C., and Garche, J. (2004) Aging mechanisms of lithium cathode materials. *Journal of Power Sources*, **127** (1–2), 58–64.

[8] Vetter, J., Novák, P., Wagner, M. *et al.* (2005) Ageing mechanisms in lithium-ion batteries. *Journal of Power Sources*, **147** (1–2), 269–281.

[9] Zhang, Y. and Wang, C.Y. (2009) Cycle-life characterization of automotive lithium-ion batteries with $LiNiO_2$ cathode. *Journal of The Electrochemical Society*, **156** (7), A527–A535.

[10] Pop, V., Bergveld, H.J., Danilov, D. *et al.* (2008) *Battery Management Systems: Accurate State-of-Charge Indication for Battery-Powered Applications*, Springer, New York, NY.

[11] Newman, J. and Thomas-Alyea, K.E. (2004) *Electrochemical Systems*, John Wiley & Sons, Inc., Hoboken, NJ.

[12] Fogler, H.S. (2005) *Elements of Chemical Reaction Engineering*, 3rd edn, Prentice Hall, Upper Saddle River, NJ.

[13] Fang, W. (2010) Fundamental modeling the performance and degradation of HEV Li-ion battery, PhD thesis, The Pennsylvania State University.

[14] Ramadass, P., Haran, B., White, R., and Popov, B.N. (2003) Mathematical modeling of the capacity fade of Li-ion cells. *Journal of Power Sources*, **123** (2), 230–240.

[15] Randall, A.V., Perkins, R.D., Zhang, X., and Plett, G.L. (2012) Controls oriented reduced order modeling of solid-electrolyte interphase layer growth. *Journal of Power Sources*, **209**, 282–288.

[16] Arpaci, V.S. (1966) *Conduction Heat Transfer*, Addison-Wesley, Reading, MA.

[17] Hill, J.M. and Dewynne, J.N. (1987) *Heat Conduction*, Blackwell Scientific Publications, Oxford.

[18] Gebhart, B. (1993) *Heat Conduction and Mass Diffusion*, McGraw-Hill, New York, NY.

[19] Jacobsen, T. and West, K. (1995) Diffusion impedance in planar, cylindrical, and spherical symmetry. *Electrochimica Acta*, **40** (2), 255–262.

[20] Forman, J.C., Bashash, S., Stein, J.L., and Fathy, H.K. (2011) Reduction of an electrochemistry-based Li-ion battery model via quasi-linearization and Padé approximation. *Journal of the Electrochemical Society*, **158** (2), A93–A101.

[21] Subramanian, V.R., Ritter, J.A., and White, R.E. (2001) Approximate solutions for galvanostatic discharge of spherical particles. *Journal of the Electrochemical Society*, **148** (11), E444–E449.

[22] Subramanian, V.R., Tapriyal, D., and White, R.E. (2004) A boundary condition for porous electrodes. *Electrochemical and Solid-State Letters*, **7** (9), A259–A263.

[23] Santhanagopalan, S., Guo, Q., Ramadass, P., and White, R.E. (2006) Review of models for predicting the cycling performance of lithium ion batteries. *Journal of Power Sources*, **156** (2), 620–628.

[24] Reddy, J.N. and Gartling, D.K. (2000) *The Finite Element Method in Heat Transfer and Fluid Dynamics*, CRC Press, Boca Raton, FL.

[25] Smith, K.A., Rahn, C.D., and Wang, C.Y. (2007) Control oriented 1D electrochemical model of lithium ion battery. *Energy Conversion and Management*, **48**, 2565–2578.

[26] Yener, Y. and Kakac, S. (2008) *Heat Conduction*, Taylor and Francis, New York, NY.

[27] Ozisik, M.N. (1980) *Heat Conduction*, John Wiley & Sons, Inc., New York, NY.

[28] Pintelon, R. and Schoukens, J. (2005) *System Identification: A Frequency Domain Approach*, IEEE Press, New York, NY.

[29] Ljung, L. (1999) *System Identification – Theory For the User*, 2nd edn, PTR Prentice Hall, Upper Saddle River, NJ.

[30] Belegundu, A. and Chandrupatla, T.R. (1999) *Optimization Concepts and Applications in Engineering*, Prentice Hall, Upper Saddle River, NJ.

[31] Nocedal, J. and Wright, S.J. (2006) *Numerical Optimization*, Springer, New York, NY.

[32] Boyd, S. and Vandenberghe, L. (2004) *Convex Optimization*, Cambridge University Press, Cambridge, UK.

[33] Hunt, G. (1996) *Electric Vehicle Battery Test Procedures Manual, Rev. 2*, US Advanced Battery Consortium.

[34] PNGV (2001) *PNGV Battery Test Manual, Rev. 3, DOE/ID-10597*, US Department of Energy.

[35] Bergveld, H.J., Kruijt, W.S., and Notten, P.H.L. (2002) *Battery Management Systems: Design by Modeling*, Kluwer Academic Publishers, Boston, MA.

[36] Gu, W.B., Wang, C.Y., and Liaw, B.Y. (1997) Numerical modeling of coupled electrochemical and transport processes in lead–acid batteries. *Journal of the Electrochemical Society*, **144** (6), 2053–2061.

[37] Barsoukov, E. and Mcdonald, J.R. (2005) *Impedance Spectroscopy: Theory, Experiment, and Applications*, John Wiley & Sons, Inc., New York, NY.

[38] Bard, A.J. and Faulkner, L.R. (2001) *Electrochemical Methods: Fundamentals and Applications*, John Wiley & Sons, Inc., New York, NY.

[39] Chen, C.T. (1999) *Linear System Theory and Design*, Oxford University Press, New York, NY.

[40] Jamshidi, M. (1983) *Large Scale Systems*, North-Holland, New York, NY.

[41] Kokotovic, P.V., Khalil, H.K., and O'Reilly, J. (1986) *Singular Perturbations in Control: Analysis and Design*, Academic Press, London.

[42] Christophides, P.D. (2001) *Nonlinear and Robust Control of PDE Systems – Methods and Applications to Transport-Reaction Processes*, Birkhauser, Boston, MA.

[43] Varga, A. (1991) Balancing-free square-root algorithm for computing singular perturbation approximations, in *IEEE Proceedings of the 30th IEEE Conference on Decision and Control*, vol. 2, IEEE Press, pp. 1062–1065.

[44] Smith, K.A., Rahn, C.D., and Wang, C.Y. (2008) Model order reduction of 1D diffusion systems via residue grouping. *ASME Journal of Dynamic Systems, Measurement, and Control*, **130** (1), 011012.

[45] Bode, H. (1977) *Lead–Acid Batteries*, Electrochemical Society Series, John Wiley & Sons.

[46] Hejabi, M., Oweisi, A., and Gharib, N. (2006) Modeling of kinetic behavior of the lead dioxide electrode in a lead–acid battery by means of electrochemical impedance spectroscopy. *Journal of Power Sources*, **158** (2), 944–948.

[47] Srinivasan, V., Wang, G.Q., and Wang, C.Y. (2003) Mathematical modeling of current-interrupt and pulse operation of valve-regulated lead acid cells. *Journal of the Electrochemical Society*, **150** (3), A316–A325.

[48] Wang, C.Y., Gu, W.B., and Liaw, B.Y. (1998) Micro-macroscopic coupled modeling of batteries and fuel cells: I. Model development. *Journal of the Electrochemical Society*, **145** (10), 3407–3417.

[49] Gu, W.B., Wang, G.Q., and Wang, C.Y. (2002) Modeling the overcharge process of VRLA batteries. *Journal of Power Sources*, **108**, 174–184.

[50] Shen, Z., Guo, J., Wang, C., and Rahn, C. (2011) Ritz model of a lead–acid battery for electric locomotives, in *ASME 2011 Dynamic Systems and Control Conference and Bath/ASME Symposium on Fluid Power and Motion Control*, vol. 1, ASME, pp. 713–720.

[51] Ramadesigan, V., Northrop, P., De, S. *et al.* (2012) Modeling and simulation of lithium-ion batteries from a systems engineering perspective. *Journal of the Electrochemical Society*, **159** (3), R31–R45.

[52] Doyle, M., Fuller, T., and Newman, J. (1993) Modeling of galvanostatic charge and discharge of the lithium/polymer/insertion cell. *Journal of the Electrochemical Society*, **140**, 1526–1533.

[53] Fuller, T., Doyle, M., and Newman, J. (1994) Simulation and optimization of the dual lithium ion insertion cell. *Journal of the Electrochemical Society*, **141**, 1–10.

[54] Paxton, B. and Newman, J. (1997) Modeling of nickel/metal hydride batteries. *Journal of the Electrochemical Society*, **144** (11), 3818–3831.

[55] Gu, W.B., Wang, C.Y., and Liaw, B.Y. (1998) Micro-macroscopic coupled modeling of batteries and fuel cells: II. Application to nickel–cadmium and nickel–metal hydride cells. *Journal of the Electrochemical Society*, **145** (10), 3418–3427.

[56] Gu, W., Wang, C., Li, S. *et al.* (1999) Modeling discharge and charge characteristics of nickel–metal hydride batteries. *Electrochimica Acta*, **44**, 4525–4541.

[57] Albertus, P., Christensen, J., and Newman, J. (2008) Modeling side reactions and nonisothermal effects in nickel metal-hydride batteries. *Journal of the Electrochemical Society*, **155** (1), A48–A60.

[58] Chaturvedi, N., Klein, R., Christensen, J. *et al.* (2010) Algorithms for advanced battery-management systems. *IEEE Control Systems Magazine*, **30** (3), 49–68.

[59] Santhanagopalan, S. and White, R. (2006) Online estimation of the state of charge of a lithium ion cell. *Journal of Power Sources*, **161**, 1346–1355.

[60] Plett, G.L. (2004) Extended Kalman filtering for battery management systems of LiPB-based HEV battery packs: Part 2. Modeling and identification. *Journal of Power Sources*, **134**, 262–276.

[61] Piller, S., Perrin, M., and Jossen, A. (2001) Methods for state-of-charge determination and their applications. *Journal of Power Sources*, **96** (1), 113–120.

[62] Pop, V., Bergveld, H.J., Notten, P.H.L., and Regtien, P.P.L. (2005) State-of-the-art of battery state-of-charge determination. *Measurement Science and Technology*, **16**, R93–R110.

[63] Plett, G.L. (2004) Extended Kalman filtering for battery management systems of LiPB-based HEV battery packs: Part 1. Background. *Journal of Power Sources*, **134**, 252–261.

[64] Plett, G.L. (2004) Extended Kalman filtering for battery management systems of LiPB-based HEV battery packs: Part 3. State and parameter estimation. *Journal of Power Sources*, **134**, 277–292.

[65] Santhanagopalan, S. and White, R. (2008) State of charge estimation for electrical vehicle batteries, in *IEEE International Conference on Control Applications*, IEEE Press, pp. 690–695.

[66] Santhanagopalan, S. and White, R.E. (2010) State of charge estimation using an unscented filter for high power lithium ion cells. *International Journal of Energy Research*, **34**, 152–163.

[67] Smith, K., Rahn, C., and Wang, C.Y. (2010) Model-based electrochemical estimation and constraint management for pulse operation of lithium ion batteries. *IEEE Transactions on Control Systems Technology*, **18** (3), 654–663.

[68] Di Domenico, D., Stefanopoulou, A., and Fiengo, G. (2010) Lithium-ion battery state of charge and critical surface charge estimation using an electrochemical model-based extended Kalman filter. *Journal of Dynamic Systems, Measurement, and Control*, **132** (6), 061302.

[69] Wang, S., Verbrugge, M., Wang, J.S., and Liu, P. (2011) Multi-parameter battery state estimator based on the adaptive and direct solution of the governing differential equations. *Journal of Power Sources*, **196** (20), 8735–8741.

[70] Schmidt, A.P., Bitzer, M., Imre, A.W., and Guzzella, L. (2010) Model-based distinction and quantification of capacity loss and rate capability fade in Li-ion batteries. *Journal of Power Sources*, **195**, 7634–7638.

[71] Safari, M., Morcrette, M., Teyssot, A., and Delacourt, C. (2010) Life-prediction methods for lithium-ion batteries derived from a fatigue approach I. Introduction: capacity-loss prediction based on damage accumulation. *Journal of the Electrochemical Society*, **157** (6), A713–A720.

[72] Spotnitz, R. (2003) Simulation of capacity fade in lithium-ion batteries. *Journal of Power Sources*, **113** (1), 72–80.

[73] Wenzl, H., Baring-Gould, I., Kaiser, R. *et al.* (2005) Life prediction of batteries for selecting the technically most suitable and cost effective battery. *Journal of Power Sources*, **144** (2), 373–384.

[74] Coleman, M., Hurley, W.G., and Lee, C.K. (2008) Improved battery characterization method using a two-pulse load test. *IEEE Transactions on Energy Conversion*, **23** (2), 708–713.

[75] Plett, G.L. (2011) Recursive approximate weighted total least squares estimation of battery cell total capacity. *Journal of Power Sources*, **196**, 2319–2331.

[76] Plett, G.L. (2011) System and method for recursively estimating battery cell total capacity. US Patent 8,041,522.

[77] Pop, V., Bergveld, H., Notten, P. *et al.* (2009) Accuracy analysis of the state-of-charge and remaining run-time determination for lithium-ion batteries. *Measurement*, **42**, 1131–1138.

[78] Åström, K.J. and Wittenmark, B. (2008) *Adaptive Control*, 2nd edn, Dover Publications, Inc., Mineola, NY.

[79] Sastry, S. and Bodson, M. (1989) *Adaptive Control: Stability, Convergence, and Robustness*, Prentice Hall, Englewood Cliffs, NJ.

[80] Goebel, K., Saha, B., Saxena, A. *et al.* (2008) Prognostics in battery health management. *IEEE Instrumentation & Measurement Magazine*, **11** (4), 33–40.

[81] Wen, J. (2009) Cell balancing buys extra run time and battery life. *Texas Instruments Analog Applications Journal*, (1Q), 14–18.

[82] PMP Portable Power (2006) Theory and implementation of Impedance Track^{TM} battery fuel-gauging algorithm in bq20zxx product family, Texas Instruments Inc., http://www.ti.com/lit/an/slua364b/slua364b.pdf.

[83] Morrow, K., Karner, D., and Francfort, J. (2008) Plug-in hybrid electric vehicle charging infrastructure review, Technical Report INL/EXT-08-15058, US Department of Energy.

[84] Ehrlich, G.M. (2002) Lithium ion batteries, in *Handbook of Batteries* (eds D. Linden and T. Reddy), 3rd edn, McGraw-Hill, New York, pp. 53–59.

[85] Brannen, N. (2008) Analysis, design and implementation of a charge-equalization circuit for use in automotive battery management systems, Master's thesis, The Pennsylvania State University.

[86] Speltino, C., Stefanopoulou, A., and Fiengo, G. (2010) Cell equalization in battery stacks through state of charge estimation polling, in *American Control Conference (ACC), 2010*, IEEE Press, pp. 5050–5055.

[87] Uno, M. and Tanaka, K. (2011) Influence of high-frequency charge–discharge cycling induced by cell voltage equalizers on the life performance of lithium-ion cells. *IEEE Transactions on Vehicular Technology*, **60** (4), 1505–1515.

[88] West, S. and Krein, P. (2000) Equalization of valve-regulated lead–acid batteries: issues and life test results, in *INTELEC. Twenty-Second International Telecommunications Energy Conference*, IEEE Press, pp. 439–446.

[89] Gu, W.B. and Wang, C. (2000) Thermal-electrochemical modeling of battery systems. *Journal of the Electrochemical Society*, **147**, 2910–2922.

[90] Ji, Y., Zhang, Y., and Wang, C. (2012) Li-ion cell operation at low temperatures, *in preparation*.

[91] Ji, Y. and Wang, C. (2012) Heating strategies for Li-ion batteries operated from subzero temperatures, *in preparation*.

[92] Luo, G., Shaffer, C., and Wang, C. (2012) Electrochemical-thermal coupled modeling for battery pack design, in *Proceedings of PRiME 2012*, Honolulu, HI.

Index

Battery Systems Engineering, First Edition. Christopher D. Rahn and Chao-Yang Wang.
© 2013 John Wiley & Sons, Ltd. Published 2013 by John Wiley & Sons, Ltd.

Printed and bound by CPI Group (UK) Ltd, Croydon, CR0 4YY

16/04/2025

14658562-0002